A TEXT BOOK OF

PHYSICS

Paper-III Core Subject (DCS 1B)

Heat and Thermodynamics

FOR

B.Sc. I : Semester - II

As Per Choice Based Credit System Syllabus (CBCS Pattern)

of Punyashlok Ahilyadevi Holkar Solapur University,

Solapur w.e.f. June 2019

Dr. R. N. MULIK
M.Sc., B.Ed., M.Phil., Ph.D.
Head, Department of Physics,
D.B.F. Dayanand College of
Arts and Science,
Solapur

Dr. S. G. HOLIKATTI
M.Sc., B.Ed., Ph.D.
Department of Physics,
Walchand College of
Arts and Science,
Solapur

Dr. S. G. PAWAR
M.Sc., Ph.D.
Department of Physics,
D.B.F. Dayanand College of
Arts and Science, Solapur

Dr. C. V. CHANMAL
M.Sc., SET, Ph.D.
Department of Physics,
D.B.F. Dayanand College of
Arts and Science, Solapur

Dr. S. D. CHAVAN
M.Sc., Ph.D.
Department of Physics,
D.B.F. Dayanand College
of Arts & Science, Solapur.

N5206

DISTRIBUTION CENTRES

PUNE

Nirali Prakashan : 119, Budhwar Peth, Jogeshwari Mandir Lane,
Pune 411002, Maharashtra.
Tel : (020) 2445 2044, Mob. No. 9657703145.
Email: bookorder@pragationline.com,
niralilocal@pragationline.com

Nirali Prakashan : S. No. 28/27, Dhyari, Near Pari Company, Pune 411041
Tel : (020) 24690204 Fax : (020) 24690316
Email : dhyari@pragationline.com,
bookorder@pragationline.com

MUMBAI

Nirali Prakashan : 385, S.V.P. Road, Rasdhara Co-op. Hsg. Society Ltd.,
Girgaum, Mumbai 400004, Maharashtra
Tel : (022) 2385 6339 / 2386 9976, Fax : (022) 2386 9976
Email : niralimumbai@pragationline.com

DISTRIBUTION BRANCHES

JALGAON

Nirali Prakashan : 34, V. V. Golani Market, Navi Peth, Jalgaon 425001,
Maharashtra, Tel : (0257) 222 0395,
Mob : 94234 91860

KOLHAPUR

Nirali Prakashan : New Mahadvar Road, Kedar Plaza, 1st Floor Opp. IDBI Bank
Kolhapur 416 012, Maharashtra. Mob : 9850046155

NAGPUR

Nirali Prakashan : Above Maratha Mandir, Shop No. 3, First Floor,
Rani Jhanshi Square, Sitabuldi,
Nagpur 440012, Maharashtra Tel : (0712) 254 7129

DELHI

Nirali Prakashan : 4593/15, Basement, Aggarwal Lane, Ansari Road,
Daryaganj, Near Times of India Building, New Delhi 110002
Mob : 08505972553

BANGALURU

Nirali Prakashan : Maitri Ground Floor, Jaya Apartments, No. 99, 6th Cross,
6th Main, Malleswaram, Bangaluru 560 003, Karnataka
Mob : +91 9449043034
Email: niralibangalore@pragationline.com

PREFACE

We, the authors are pleased to present this text book of Physics **(Heat and Thermodynamics)** for B.Sc. I, Punyashlok Ahilyadevi Holkar Solapur University, Solapur, for Semester II in the market and care, therefore, has been taken, while preparing the text book, that it caters the needs not only of the students and teachers concerned, but it also creates interest and inquisitiveness about the subject in any person, who lays hand on it.

This book has been written strictly according to the guidelines of the revised syllabus (CBCS Pattern) of B.Sc. I Physics (Heat and Thermodynamics) prescribed by Board of Studies in Physics, Punyashlok Ahilyadevi Holkar Solapur University, Solapur. It is our humble belief that the text books are among the invaluable resources for successful teaching – learning process, provided they are written within the framework of the aims and objectives laid down and hence this book is really the honest efforts in that direction.

Much more attention has been paid in simplifying the topics and presenting them in a simple and scientific language. Each concept is lucidly explained and supported by self-explanatory diagrams. A few examination oriented problems have been solved in every topic and few are given for practice. The key statements, laws, principles and definitions are printed out from the rest of the matter. As per the nature of university question paper, number of multiple choice questions, short and long answer type questions are included for self testing. Every attempt has been made to make the matter easily readable and readily understandable. In fact, the book has been written out of a prolonged teaching experience.

Every care has been taken to check the mistakes and misprints, yet it is very difficult to claim perfection. Any errors, omission and suggestions for the improvement of this text book, if brought to our notice will be thankfully acknowledged and incorporated in the next edition.

Authors are very much thankful to Shri. Dineshbhai Furia of Nirali Prakashan whose inspiration and constant prompting are responsible for producing this text book in short period. Authors are also very much thankful to Shri. Jignesh Furia who have contributed quite a lot for this publication. We reserve special thanks to Shri. Prabhakar D. Nandkile and Shri. Kiran Velankar for active help throughout this work. In fact entire staff of Nirali Prakashan especially Mr. Santosh Bare and Mrs. Prachi Sawant has put in a lot of efforts for the publication of this book. Authors are thankful to all those who have contributed and helped during compilation of this book.

We hope that this text book will be useful to the teachers and students preparing for B.Sc. I examination of Punyashlok Ahilyadevi Holkar Solapur University, Solapur. Also we hope this book will receive spontaneous response from teachers and students.

Authors

SYLLABUS

---

1. TRANSPORT PHENOMENON (08)

Introduction, Mean free path, Clausius expression for mean free path (Collision cross-section), Transport phenomenon, Coefficient of viscosity, Thermal conductivity and its dependence on temperature and pressure.

2. LIQUEFACTION OF GASES (08)

Liquefaction of gases by Joule-Thomson effect, Linde's air liquefier, Cooling by adiabatic demagnetization and expression for fall in temperature, Experimental set-up for adiabatic demagnetization of paramagnetic substances, Properties of liquid helium.

3. THERMODYNAMICS (10)

Laws of thermodynamics, Reversible and Irreversible processes, Isothermal and Adiabatic process, Adiabatic relations, Work done during isothermal and adiabatic processes, Entropy change in reversible and irreversible processes.

4. HEAT ENGINES (08)

Introduction, Carnot's heat engine and its efficiency, Heat engine, Otto cycle and its efficiency, Diesel cycle and its efficiency, Comparison between Otto and Diesel engine.

5. REFRIGERATOR (08)

General principle, Refrigeration cycle, Coefficient of performance of refrigerator, Vapour compression refrigerator, Air conditioning (principle and applications).

□□□

CONTENTS

•••

Punyashlok Ahilyadevi Holkar Solapur University, Solapur
Nature of Question Paper for Choice Based Credit System
(CBCS) Semester Pattern
• Faculty of Science •
(w.e.f. June 2019)

Time : 2 Hrs.   **Total Marks 40**

Instructions :

1.   All questions are compulsory.
2.   Draw neat diagrams and give equations wherever necessary.
3.   Figures to the right indicate full marks.
4.   Use of logarithmic table and calculator is allowed.

Q. No. 1 Multiple Choice Questions :   (08)
1.   ....
    (a) .... (b) .... (c) .... (d) ....
2.   ...do...
3.   ...do...
4.   ...do...
5.   ...do...
6.   ...do...
7.   ...do...
8.   ...do...

Q. No. 2 Answer any Four of the following :   (08)
(i)
(ii)
(iii)
(iv)
(v)
(vi)

Q. No. 3 (A) Write notes on any One of the following :   (03)
(i)
(ii)

    (B) Solve/Short answer :   (05)

Q. No. 4 Answer any Two of the following :   (08)
(i)
(ii)
(iii)

Q. No. 5 Answer any One of the following :   (08)
(i)
(ii)

•••

1

CHAPTER

TRANSPORT PHENOMENA

1.1 INTRODUCTION

According to the kinetic theory of gases the molecules of a gas are almost free to move and have random motion. They collide on each other. During motion, a gas molecule carries certain physical quantities such as momentum, energy, mass etc.

When the gas is not in equilibrium the random motion of gas molecules gives rise to viscosity, conductivity and diffusion. Under this condition the gas tries to attain an equilibrium state by transporting momentum, energy and mass.

1.2 MEAN FREE PATH

Due to random motion the gas molecules suffer continuous collisions with one another. Between any two successive collisions, a molecule travels a straight line path. At every collision the direction of molecule changes. Thus the path of a single molecule consists of a series of short zig-zag paths.

The straight line path covered by a gas molecule between any two successive collisions is known as the free path.

The path of a single molecule is as shown in Fig. 1.1.

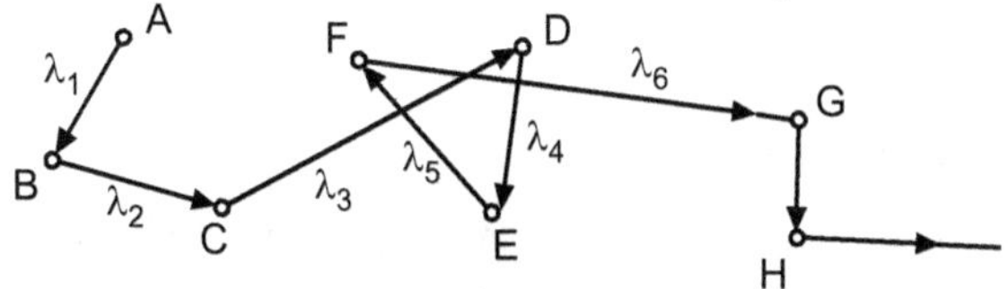

Fig. 1.1

It is seen that after every collision the direction of a molecule changes. We see from above figure that the free paths of a molecule are λ_1, λ_2, λ_3, ... Some paths are long and some are short. Therefore, average or mean free path is defined. *'Mean free path is defined as the average distance travelled by a molecule between two successive collisions'.*

(1.1)

Let λ_1, λ_2, λ_3, ..., λ_n be the free paths of a molecule suffering n collisions then, mean free path,

$$\lambda = \frac{\lambda_1 + \lambda_2 + \lambda_3 + ... + \lambda_n}{n} \qquad \qquad ... (1.1)$$

Sometimes mean free path is denoted by $\bar{\lambda}$ or λ_{av}.

1.3 CLAUSIUS EXPRESSION FOR MEAN FREE PATH (COLLISION CROSS SECTION)

Collision cross section :

Let us assume that a molecule, say A of a gas is moving and all others are at rest as shown in Fig. 1.2.

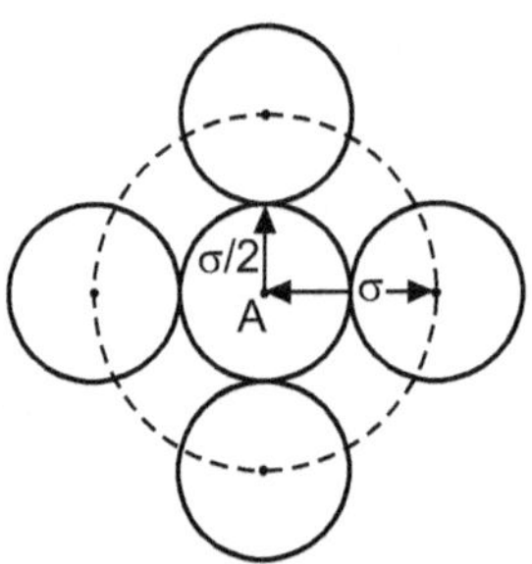

Fig. 1.2

Let σ be the diameter of each molecule. The sphere of radius σ with centre of molecule A is the sphere of influence. Then the cross-section $\pi\sigma^2$ along the diameter of a sphere of influence is a collision cross-section.

Expression for mean free path :

To determine the mean free path, it is assumed that (i) only one molecule is moving and all others are at rest and (ii) the sphere of influence of each molecule has a radius σ (here σ is the diameter of each molecule).

The mass of the gas is assumed to be divided into parallel layers. Let the thickness of each layer be equal to diameter of a single molecule, σ. Let n be the number of molecules per unit volume (i.e. molecular density) of the gas.

Let a molecule A move with r.m.s. velocity c through the mass as shown in Fig. 1.3.

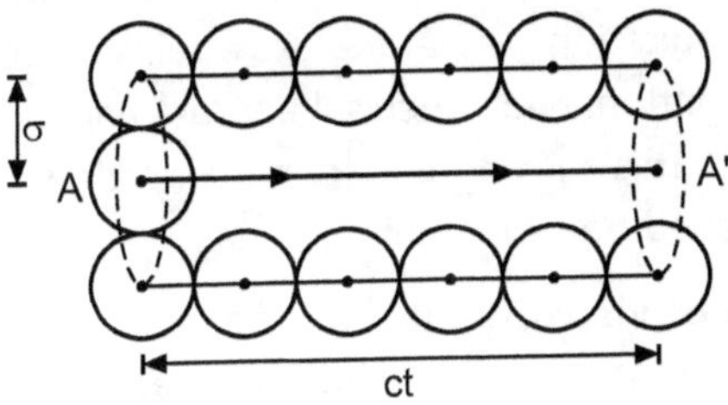

Fig. 1.3

The molecule A will collide with all the molecules whose centres lie within a cylinder of radius σ, while travelling along the line AA'. Here $\pi\sigma^2$ is the mutual collision cross-section of the molecule A and other molecules.

The molecule covers a distance ct in time t. Thus the length of the cylinder will be AA' = ct. Then the volume of the cylinder $V = \pi\sigma^2 ct$.

$\therefore$ Number of molecules in the cylinder, $N = \pi\sigma^2 ctn$.

$\therefore$ Number of collisions by molecule A along the path AA' is also,

$$N = \pi\sigma^2 ctn$$

$\therefore$ The mean free path,

$$\lambda = \frac{\text{Total distance covered}}{\text{Number of collisions}}$$

$$= \frac{ct}{N}$$

$$= \frac{ct}{\pi\sigma^2 ctn}$$

$\therefore$ $$\lambda = \frac{1}{\pi\sigma^2 n} \qquad \text{... (1.2)}$$

This is the Clausius expression for mean free path. It is seen from the above equation that the mean free path is inversely proportional to the molecular density of the gas.

1.4 TRANSPORT PHENOMENA

According to the kinetic theory of gases, the molecules of a gas are in a state of thermal agitation. Therefore, gas attains an equilibrium state by transporting momentum, heat (thermal energy) and mass from one layer to another layer. This gives rise to the viscosity, conductivity and diffusion. This is called transport phenomena.

Phenomenon of viscosity : It is due to the transport of momentum. This is because different parts of the gas may have different velocities.

Phenomenon of thermal conductivity : It is due to the transport of heat (kinetic energy). This is because the temperature of the gas may not be uniform.

Phenomenon of diffusion : It is due to the transport of mass of a gas. This is because different parts of the gas may have different concentrations.

1.5 COEFFICIENT OF VISCOSITY

The different layers of a gas may have different velocities. This will result in relative motion between the layers. The layer moving faster will impart momentum on layer moving slower. This gives rise to the phenomenon of viscosity of the gas.

Consider a gas flowing over a horizontal surface as shown in Fig. 1.4.

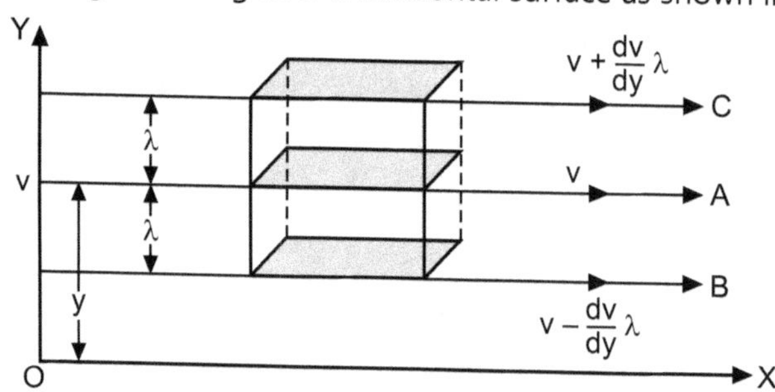

Fig. 1.4

The velocity of the layer in contact with the surface OX is zero. The velocity increases along vertical direction at a uniform rate $\frac{dv}{dy}$. Here $\frac{dv}{dy}$ is known as velocity gradient.

Now consider a layer A moving with velocity v. It is at a distance y from OX. All the molecules in this layer have the same velocity.

Consider two more layers B and C just below and above A respectively at a distance λ. Here, λ is equal to mean free path of the molecules.

The velocity of a gas in the layer C $= v + \dfrac{dv}{dy}\lambda$ and the velocity of the gas in the layer B $= v - \dfrac{dv}{dy}\lambda$.

Due to thermal agitation, the gas molecules are moving in all directions. Therefore, the average number of molecules moving along any one (+ve or −ve) axis will be $\dfrac{1}{6}^{\text{th}}$ of the total number of molecules in the gas.

Let n be the number of molecules per unit volume, m be the mass of each gas molecule and c be the average velocity of a molecule at a given temperature of the gas.

Then the number of molecules passing from C to B downwards or B to C upwards per unit area per second is equal to $\dfrac{nc}{6}$.

$\therefore$ The momentum carried by the molecules from layer C to layer A, per unit area per second $= m \times \dfrac{nc}{6} \times \left(v + \dfrac{dv}{dy}\lambda\right)$ ... (1.3)

Also the momentum carried by the molecules from layer B to layer A, per unit area per second $= m \times \dfrac{nc}{6} \times \left(v - \dfrac{dv}{dy}\lambda\right)$ ... (1.4)

Therefore, net momentum transferred per unit area per second at layer A

$$= \dfrac{mnc}{6}\left\{\left(v + \dfrac{dv}{dy}\lambda\right) - \left(v - \dfrac{dv}{dy}\lambda\right)\right\}$$

$$= \dfrac{mnc}{6} \times 2\dfrac{dv}{dy}\lambda$$

$$= \dfrac{1}{3}mnc\lambda\dfrac{dv}{dy} \qquad\qquad ... (1.5)$$

This is the tangential force per unit area.

Now, if η is the coefficient of viscosity of the gas and for velocity gradient $\dfrac{dv}{dy}$, the tangential force (viscous force) per unit area is $\eta\,\dfrac{dv}{dy}$.

$\therefore$ In steady state,

$$\eta\,\frac{dv}{dy} = \frac{1}{3}\,mnc\lambda\,\frac{dv}{dy}$$

$\therefore$ $$\eta = \frac{1}{3}\,mnc\lambda \qquad\qquad \dots (1.6)$$

$\therefore$ $$\boxed{\eta = \frac{1}{3}\,\rho c\lambda} \qquad\qquad \dots (1.7)$$

where $\rho = mn$, density of the gas.

1.5.1 Dependence of η on Temperature

We have, $$\eta = \frac{1}{3}\,\rho c\lambda$$

The density (ρ) of the gas decreases with increase in temperature but the mean free path λ increases in the same proportion. Thus $\rho\lambda$ remains constant. Since the average velocity (c) is directly proportional to the square root of absolute temperature ($c \propto \sqrt{T}$), the coefficient of viscosity (η) will also be proportional to $\sqrt{T}$.

$\therefore$ $$\eta \propto \sqrt{T}$$

Thus the coefficient of viscosity of a gas is directly proportional to the square root of its absolute temperature.

1.5.2 Dependence of η on Pressure

The density (ρ) of a gas increases with increase in pressure but the mean free path (λ) decreases in the same proportion. Thus $\rho\lambda$ remains constant. Also the average velocity (c) is independent of pressure. Thus, the coefficient of viscosity, $\eta\left(= \dfrac{1}{3}\,\rho c\lambda\right)$ is independent of pressure.

1.6 THERMAL CONDUCTIVITY

The different layers of a gas may have different temperatures. The molecules may carry heat (kinetic energy) from region of high temperature to the region of low temperature. This gives rise to the phenomenon of thermal conductivity.

Consider a fixed mass of a gas at rest as shown in Fig. 1.5. Suppose there is a uniform temperature gradient $\dfrac{d\theta}{dy}$ along vertical direction (along OY). Temperature of each horizontal layer along OX is constant.

Let θ be the temperature of layer A. The temperature of layer C at a distance λ above layer A is $\left(\theta + \dfrac{d\theta}{dy}\lambda\right)$. The temperature of layer B at a distance λ below layer A is $\left(\theta - \dfrac{d\theta}{dy}\lambda\right)$. Here λ is the mean free path.

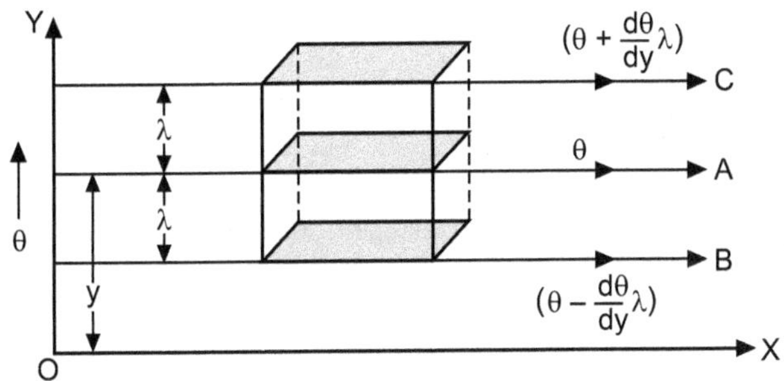

Fig. 1.5

Thus the gas above layer A is at high temperature than the gas below layer A. Molecules coming from C downwards passing across A possess more heat (K.E.) than the molecules coming from B upwards across A.

Let n be the number of molecules per unit volume, m be the mass of each gas molecule and c be the average velocity of molecule at a given temperature.

The number of molecules crossing unit area of the layer A upwards or downwards per second is $\dfrac{nc}{6}$.

If C_v be the specific heat of the gas at constant volume, then the heat energy carried by the molecules in crossing unit area of layer A in downward direction per second

$$= \text{Mass} \times \text{Specific heat} \times \text{Temperature}$$

$$= \frac{mnc}{6} \times C_v \times \left(\theta + \frac{d\theta}{dy}\lambda\right)$$

Similarly, heat energy carried by molecules crossing unit area of layer A in upward direction per second

$$= \frac{mnc}{6} \times C_v \times \left(\theta - \frac{d\theta}{dy}\lambda\right)$$

$\therefore$　Net transfer of heat energy per unit area per second across the layer A,

$$Q = \frac{mnc}{6} \times C_v \times \left(\theta + \frac{d\theta}{dy}\lambda\right) - \frac{mnc}{6} \times C_v \times \left(\theta - \frac{d\theta}{dy}\lambda\right)$$

$$= \frac{mnc}{6} \times C_v \times 2\lambda \frac{d\theta}{dy}$$

$$= \frac{1}{3} mnc \times C_v \lambda \frac{d\theta}{dy} \qquad \ldots (1.8)$$

Now, the coefficient of thermal conductivity (k) of the gas is defined as the quantity of heat that flows per unit area per second per unit temperature gradient.

$$\therefore \qquad Q = k\frac{d\theta}{dy} \qquad \ldots (1.9)$$

$\therefore$　Equating equations (1.8) and (1.9), we get

$$k\frac{d\theta}{dy} = \frac{1}{3} mnc\, C_v \lambda \frac{d\theta}{dy}$$

$$\therefore \qquad k = \frac{1}{3} mnc\, C_v \lambda \qquad \ldots (1.10)$$

Since, mn = ρ, density of the gas,

$$k = \frac{1}{3} \rho c\, C_v \lambda \qquad \ldots (1.11)$$

Also, $\qquad\qquad k = \eta C_v$ (from equation (1.7)

1.6.1 Dependence of k on Temperature

We have, $k = \dfrac{1}{3} \rho c \, C_v \lambda$

The density (ρ) decreases with increase in temperature but mean free path λ increases in such a way that $\rho\lambda$ remains constant. Since C_v is very small and $c \propto \sqrt{T}$, $k \propto \sqrt{T}$. Thus the coefficient of thermal conductivity is directly proportional to the square root of absolute temperature.

1.6.2 Dependence of k on Pressure

As $\rho\lambda$ remains constant for change in pressure and c is independent of pressure, the coefficient of thermal conductivity is independent of pressure.

SOLVED PROBLEMS

Problem 1.1 : *The diameter of atom of a gas molecule is 0.9 A°. One mole of gas occupies 10 litres at 20°K. Calculate the mean free path of the molecules. (Avogadro's number, $N_A = 6 \times 10^{23}$ mole^{-1}).*

Given : Volume, $V = 10$ litres $= 10 \times 10^{-3}$ m^3

$$N_A = 6 \times 10^{23} \text{ mole}^{-1}$$

$$\sigma = 0.9 \text{ A}° = 0.9 \times 10^{-10} \text{ m}$$

Solution : $n = \dfrac{N_A}{V} = \dfrac{6 \times 10^{23}}{10 \times 10^{-3}} = 6 \times 10^{25}$ m^{-3}

$\therefore$ $\lambda = \dfrac{1}{\pi\sigma^2 n} = \dfrac{1}{3.14 \times (0.9 \times 10^{-10})^2 \times 6 \times 10^{25}}$

$$= \mathbf{6.553 \times 10^{-7} \text{ m}}$$

Problem 1.2 : *Calculate the coefficient of viscosity of a gas having average velocity of 900 m/s. (Density of gas = 1.5 kg/m^3 and mean free path = 8×10^{-6} m)*

Given : $c = 900$ m/s, $\rho = 1.5$ kg/m^3, $\lambda = 8 \times 10^{-6}$ m

Solution : $\eta = \dfrac{1}{3} \rho c \lambda$

$$= \dfrac{1}{3} \times 1.5 \times 900 \times 8 \times 10^{-6}$$

$$= \mathbf{3.6 \times 10^{-3} \text{ N.s/m}^2}$$

Or $\eta = \mathbf{3.6 \times 10^{-3} \text{ kg/m.s}}$

Problem 1.3 : *Calculate the thermal conductivity of a gas having specific heat at constant volume, $C_v = 2.5 \times 10^3$ J/kg °K and coefficient of viscosity, $\eta = 1.6 \times 10^{-5}$ N.s/m^2.*

Solution :
$$k = \frac{1}{3}\rho c C_v \lambda = \eta C_v$$

$$\therefore \quad k = 1.6 \times 10^{-5} \times 2.5 \times 10^3$$

$$= \mathbf{4 \times 10^{-2} \ J/m.s \ °K}$$

EXERCISES

(A) Multiple Choice Questions :

1. Viscosity of a gas is due to transport of ......

    (a) momentum
    (b) energy
    (c) mass
    (d) none of these

2. At a given temperature, the coefficient of viscosity of a gas ......

    (a) decreases with decrease of pressure

    (b) increases with increase of pressure

    (c) is independent of pressure

    (d) is equal to pressure

3. The mean free path of gas molecules is inversely proportional to ......

    (a) square of the diameter of the molecule

    (b) square root of the diameter of the molecule

    (c) molecular diameter

    (d) fourth power of the molecular diameter

4. Viscosity of a gas is directly proportional to ......

    (a) temperature
    (b) pressure
    (c) $\sqrt{T}$
    (d) T^2

5. When the pressure of a gas increases, the mean free path of the molecules ......

    (a) increases
    (b) decreases
    (c) remains constant
    (d) none of these

6. Transport of ...... gives rise to the phenomenon of thermal conductivity of a gas.

   (a) mass　　　　　　　　　　　(b) energy

   (c) momentum　　　　　　　　(d) any one of these

7. Coefficient of thermal conductivity depends on temperature as ......

   (a) $k \propto \sqrt{T}$　　　　　　　　(b) $k \propto T$

   (c) $k \propto T^2$　　　　　　　　(d) $k \propto \dfrac{1}{T}$

8. Coefficient of thermal conductivity of a gas is related to the coefficient of viscosity as ......

   (a) $k = \dfrac{C_v}{\eta}$　　　　　　　　(b) $k = \dfrac{\eta}{C_v}$

   (c) $k = \eta\, C_v$　　　　　　　　(d) $k = \eta \rho C_v$

9. As the temperature of a gas increases, mean free path of gas molecules ......

   (a) decreases　　　　　　　　(b) remains constant

   (c) increases　　　　　　　　(d) none of these

ANSWERS

1. (a)	2. (c)	3. (a)	4. (c)	5. (b)	6. (b)	7. (a)	8. (c)	9. (c)

(B) Short Answer Type Questions :

1. Define the terms (a) free path, (b) mean free path.

2. What do you mean by collision cross-section ?

3. How viscosity of a gas depends upon temperature and pressure ?

4. How thermal conductivity of a gas depends upon temperature and pressure ?

5. What do you mean by transport phenomena in case of a gas ?

(C) Long Answer Type Questions :

1. What is mean free path ? Obtain Clausius equation for mean free path by collision cross-section method.

2. Obtain an expression for coefficient of viscosity of a gas.

3. Obtain an expression for coefficient of thermal conductivity of a gas.

4. Discuss the effect of temperature and pressure on (a) viscosity of a gas, (b) thermal conductivity of a gas.

(D) Problems for Practice :

1. One mole of He gas occupies 40 litres at 20°K. If the diameter of He atom is 1°A, calculate the mean free path of the molecules.

   (Given : Avogadro's number = 6×10^{23} mole^{-1})　　(**Ans.** 2.12×10^{-6} m)

2. Determine the mean free path of gas molecules having average velocity of 650 m/s. (Given : Density of gas = 2 kg/m^3 and coefficient of viscosity of gas = 2.5×10^{-4} N.s/m^2)　　　　(**Ans.** 5.77×10^{-7} m)

3. Calculate the specific heat of a gas at constant volume. The gas has thermal conductivity of 0.5 J/ms °K and coefficient of viscosity 2×10^{-5} N.s/m^2.　　　　　　(**Ans.** 2.5×10^{4} J/kg °K)

2

CHAPTER

LIQUEFACTION OF GASES

2.1 INTRODUCTION

The low temperature physics is one of the most important branches of physics. The production of low temperature is connected with liquefaction of gases. The first step to liquefy the air with the help of Joule-Thomson expansion effect was formerly called the porous plug experiment.

2.2 LIQUEFACTION OF GASES BY JOULE-THOMSON EFFECT

Joule-Thomson effect :

When a gas is allowed to escape adiabatically through porous plug, from a region of constant high pressure to the region of constant low pressure, it undergoes change in temperature. This phenomenon is called Joule-Thomson effect or Joule-Thomson adiabatic throttling.

Observations of Joule-Thomson effect :

(i) At ordinary temperature due to Joule-Thomson effect, all gases show change in temperature (except hydrogen and helium).

(ii) At sufficiently low temperature, all gases show cooling.

(iii) Fall in temperature is proportional to the difference in pressure on two sides of porous plug.

For most of gases, it is seen that the Joule-Thomson cooling observed is very small e.g. at temperature 20°C, when pressure difference between two sides of porous plug is 50 atmosphere then temperature falls by 11.7°C. For 210 atmosphere pressure difference, the decrease in temperature is 42°C. The cooling produced by Joule-Thomson effect depends on initial temperature of gas and difference in pressure on two sides of porous plug. However, the fall in temperature cannot be increased simply by increasing the pressure difference.

(2.1)

The cooling effect can be increased by the process of regenerative cooling. In regenerative cooling, a portion of the gas suffers Joule-Thomson expansion and becomes cooled and it is used to cool another portion of the incoming gas. The temperature of such a gas further decreases. This gas is again made to circulate and cool the incoming gas. In this way, the cooling effect becomes cumulative. This is the principle of regenerative cooling.

2.3 LINDE'S AIR LIQUEFIER

In 1896, Linde first liquefied atmospheric air using the principle of Joule-Thomson effect and regenerative cooling. The Linde's apparatus is as shown in Fig. 2.1.

Principle : Cooling the incoming gas by the gas which has already undergone cooling.

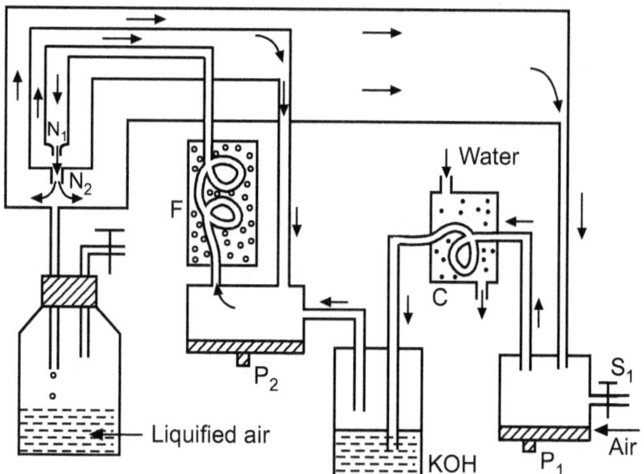

Fig. 2.1 : Linde's air liquefier

The atmospheric air is made dust free and water vapours are removed. The air is then passed through an opening S_1. The compressor P_1 applies high pressure of about 20-25 atmospheres. The air is then passed through cold water bath C which again passes through KOH solution. In KOH solution, there is absorption of water vapour and CO_2 gas. The purified air now enters into the second compressor P_2. In compressor P_2, the pressure is raised to about 200 atmospheres. The compressed air now passes through conductive cooling spiral F which is immersed in a mixture F. In F, the temperature of air is brought down to approximately – 20°C then the air comes to the first nozzle N_1, here air expands adiabatically. For first cycle, air is not liquefied, therefore, it undergoes all the steps again as shown by arrow ($\rightarrow$) marks. After many cycles, the air is finally liquefied. Linde obtained liquefied air at the rate of 1 lit/hour.

2.4 COOLING BY ADIABATIC DEMAGNETISATION OR MAGNETO CALORIC EFFECT

In a unmagnetised state, the atomic magnetic dipoles are randomly oriented. When a paramagnetic substance is placed in a magnetic field, the substance becomes magnetised as shown in Fig. 2.2 (b).

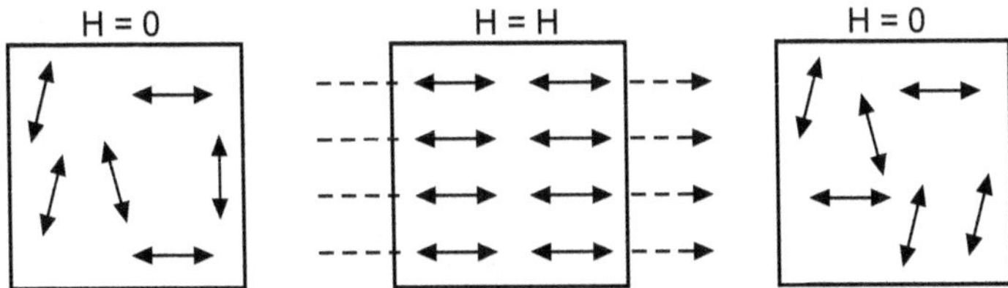

(a) Unmagnetised state (b) Magnetised state (c) Demagnetised state

Fig. 2.2

During magnetisation, work is done on the atomic magnetic dipoles. This work appears in the form of heat and temperature of the substance increases.

When external magnetic field is removed, again dipoles get randomly oriented and the and the paramagnetic substance demagnetises. The substance looses its magnetism. During demagnetisation if the substance is thermally isolated then work is done by the substance.

Energy for this is derived from the substance itself. Thus internal energy decreases resulting in fall in temperature. This change in temperature is known as magneto caloric effect. This is the principle of cooling by adiabatic demagnetisation. This is suggested theoretically by Debye and Giauque.

2.4.1 Expression for Fall in Temperature

We know the first law of thermodynamics.

$$dQ = dU + dW$$

$$\therefore \quad dQ = dU + PdV$$

But $\quad dQ = TdS$

$$\therefore \quad TdS = dU + PdV$$

$$\therefore \quad dU = TdS - PdV \qquad \ldots (2.1)$$

If we consider associated effect to such a thermodynamic system then equation (2.1) can be modified as

$$dU = TdS - PdV + HdM \qquad \ldots (2.2)$$

where H is applied magnetic field and

M is intensity of magnetisation.

When a substance is magnetised there is very small change in its volume (dV = 0).

$\therefore$　Equation (2.2) becomes

$$dU = TdS + HdM \qquad \ldots (2.3)$$

Equation (2.3) is of the type

$$dZ = M\,dx + N\,dy$$

It satisfies the below relation :

$$\left(\frac{\partial M}{\partial y}\right)_x = \left(\frac{\partial N}{\partial x}\right)_y$$

$\therefore$　Equation (2.3) becomes

$$\left(\frac{\partial T}{\partial M}\right)_S = \left(\frac{\partial H}{\partial S}\right)_M \qquad \ldots (2.4)$$

Rearranging the terms, we get

$$\left(\frac{\partial T}{\partial H}\right)_S = \left(\frac{\partial M}{\partial S}\right)_M \qquad \ldots (2.5)$$

For paramagnetic materials, as magnetic field H is increased then the intensity of magnetisation M decreases i.e. there is inverse relation between H and M.

$\therefore$ In equation (2.5) instead of constant M, we write constant H with a negative sign.

$\therefore$
$$\left(\frac{\partial T}{\partial H}\right)_S = -\left(\frac{\partial M}{\partial S}\right)_H$$

$\therefore$
$$\left(\frac{\partial T}{\partial H}\right)_S = \frac{-\left(\frac{\partial M}{\partial T}\right)_H}{\left(\frac{\partial S}{\partial T}\right)_H}$$

Multiply and divide by T,

$$\left(\frac{\partial T}{\partial H}\right)_S = \frac{-T\left(\frac{\partial M}{\partial T}\right)_H}{T\left(\frac{\partial S}{\partial T}\right)_H}$$

But $C_H = T\left(\frac{\partial S}{\partial T}\right)_H$ = Specific heat at constant magnetic field H

$\therefore$
$$\left(\frac{\partial T}{\partial H}\right)_S = \frac{-T}{C_H}\left(\frac{\partial M}{\partial T}\right)_H \qquad \ldots (2.6)$$

Since the process is carried out adiabatically, entropy S remains unchanged.

$\therefore$ Equation (2.6) becomes,

$$dT = -\frac{T}{C_H}\left(\frac{\partial M}{\partial T}\right)_H dH \qquad \ldots (2.7)$$

When magnetic field changes from H_1 to H_2, change in temperature is,

$$\Delta T = \frac{-T}{C_H}\int_{H_1}^{H_2}\left(\frac{\partial M}{\partial T}\right)_H dH \qquad \ldots (2.8)$$

But
$$M = CV\left(\frac{H}{T}\right)$$

where C is Curie temperature.

$\therefore$
$$\left(\frac{\partial M}{\partial T}\right)_H = -\frac{CVH}{T^2}$$

Substituting in equation (2.8), we get

$$\Delta T = -\frac{T}{C_H} \int_{H_1}^{H_2} \left(-\frac{CVH}{T^2}\right) dH = \frac{CV}{C_H T} \int_{H_1}^{H_2} H \, dH$$

$$\therefore \quad \boxed{\Delta T = \frac{CV}{2C_H T}(H_2^2 - H_1^2)} \qquad \dots (2.9)$$

Equation (2.9) gives the expression for fall in temperature. It also represents magneto calorie effect. This effect is used to obtain very low temperature.

2.5 EXPERIMENTAL SET UP FOR ADIABATIC DEMAGNETISATION OF PARAMAGNETIC SUBSTANCE

The experimental set up to produce low temperature by adiabatic demagnetisation of a paramagnetic salt is as shown in Fig. 2.3.

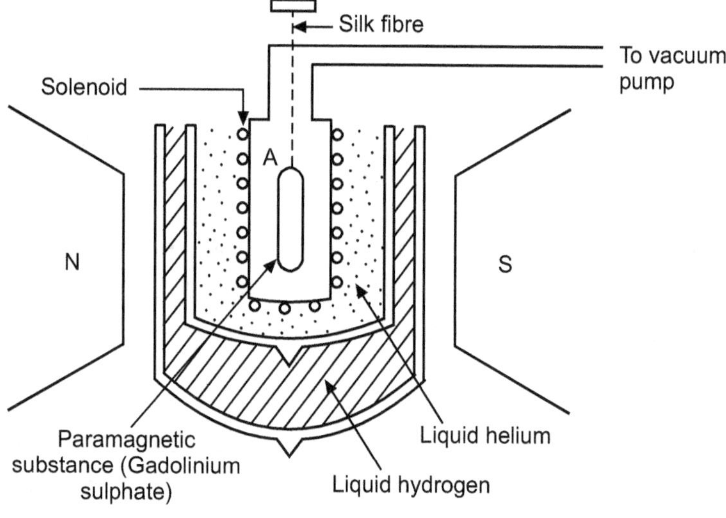

Fig. 2.3

The apparatus contains two special types of flasks called Dewar flasks. In the outer flask, liquid hydrogen is placed at temperature of 18°K and in the inner flask liquid helium is placed at 4°K. The paramagnetic substance suspended by a nylon or silk fibre in the cylindrical chamber is as shown in Fig. 2.3. This is connected to a vacuum pump. The vacuum pump is used to remove helium gas present in the cylindrical chamber (A). The outer flask is surrounded by two pole pieces N and S, which can produce very fine magnetic field of the order of 30,000 gauss.

When the strong magnetic field is applied, the substance becomes magnetised, as a result the temperature increases but this heat is absorbed by introduction of hydrogen in A and pumping it off using vacuum pump. Thus temperature of the specimen is maintained at 4°K. When the magnetic field is switched off, the substance gets demagnetised and the temperature falls adiabatically to very low value about 1°K.

Such low temperature can be measured by using a special thermometer called magnetic thermometer. The solenoid is used to measure self inductance which is used to calculate magnetic susceptibility of the substance at the beginning and at the end of experiment. The temperature T is calculated by Curie's law, $\chi = \dfrac{C}{T}$.

The temperature of the order of 0.0014 °K can be produced by this method.

2.6 PROPERTIES OF LIQUID HELIUM

K. Onnes found that when liquid helium is cooled at about 2.19°K it shows discontinuity in its density. As temperature falls from 4.2°K to 2.19°K, density increases then with further decrease in temperature, density decreases.

Kapitza observed that at 2.19°K, the specific heat C_v increases rapidly. This can be shown in Fig. 2.4.

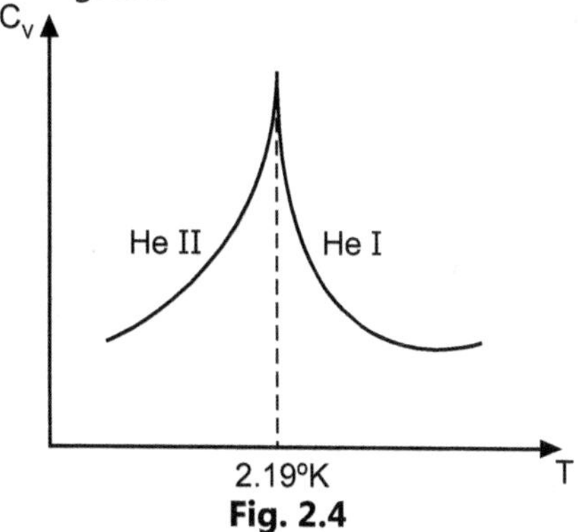

Fig. 2.4

The nature of the curve look like lambda (λ). The transition temperature is 2.19°K and is called lambda point. Liquid helium above 2.19°K is called helium I and the liquid helium below 2.19°K is called helium II. Liquid helium I behaves like an ordinary liquid but liquid helium II shows abnormal properties.

1. Superfluidity :

For an ordinary liquid as temperature decreases, viscosity increases, but for liquid helium II, viscosity decreases with decrease in temperature.

At 2.19°K the viscosity of liquid helium II becomes zero. Due to that liquid helium II can flow very easily through very small capillary tube, which is not easy for ordinary liquid. Therefore liquid helium II is called superfluid. And this extraordinary property is called superfluidity.

2. Fountain effect :

Liquid He II exhibits a property of superfluidity in the form of fountain effect as shown in Fig. 2.5.

A tube having a shape of an ink dropper with a very small opening at its top and bottom is suspended in liquid He II contained in Dewar flask as shown in Fig. 2.5.

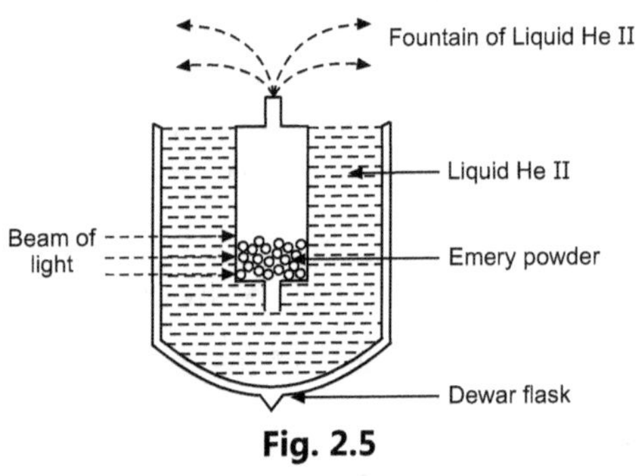

Fig. 2.5

Inside the tube, emery power is filled. The interspacing between emery fine particles provide fine capillaries for the liquid He II to flow in the tube. When the heat radiation from a light torch fall on the system containing liquid He II produces a pressure difference. The liquid helium II starts coming from the top end of the tube in the form of fountain, which is even of height 30 cm. This phenomenon is due to the absorption of heat energy by fine emery power and the capillary action of the fine emery particles.

3. Mechano-caloric effect :

This effect is opposite to that of fountain effect. This effect gives the relation between mechanical moment and temperature. The apparatus is as shown in Fig. 2.6.

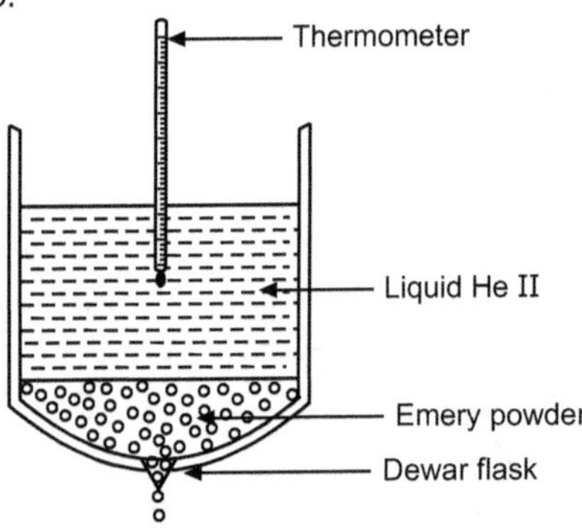

Fig. 2.6

A sensitive thermometer is dipped in liquid He II contained in Dewar flask, which has a small hole at the bottom. An emery powder is kept at the bottom of Dewar flask and above that liquid He II is held. If the drop of liquid He II is come out from the hole, the temperature raises by about 0.1°C.

4. Creeping film experiment (Rolling film) :

The experiment shows that liquid He II has very small surface tension. This was demonstrated by Rolling and hence it is called Rolling (creeping) experiment.

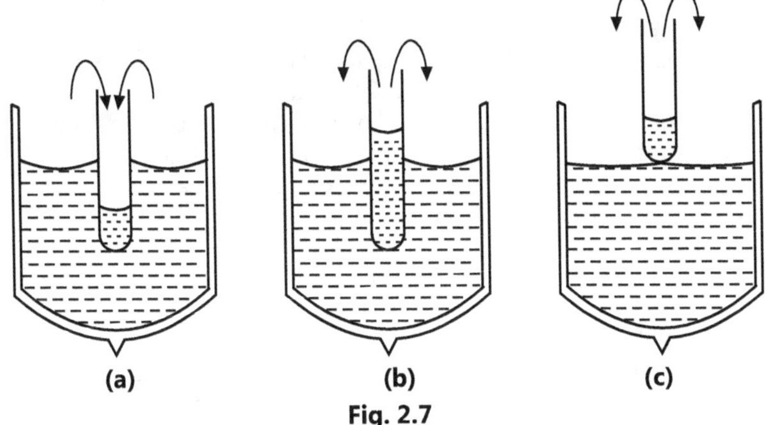

(a)   (b)   (c)

Fig. 2.7

If a tube containing liquid He II is placed in a Dewar flask containing the same liquid, following observations are seen.

(a) If the level of the liquid tube is lower than outside liquid then the liquid outside starts creeping into the tube [Refer Fig. 2.7 (a)].

(b) If the tube is raised up so that level in the tube is higher than outside it, then the liquid which is inside the tube starts creeping out of the tube [Refer Fig. 2.7 (b)], until the two levels become the same.

(c) If the tube is lifted up above the liquid surface, the liquid creeping out along the inner wall of the tube and falls into the liquid [Refer Fig. 2.7 (c)]. This is continued till entire liquid is drained out.

5. High heat conduction :

The heat conductivity of liquid He I is normal, but liquid He II has high conductivity about 800 times more than copper.

EXERCISES

(A) Multiple Choice Questions :

1. In Joule-Thomson cooling the gas allowed to escape through porous plug is from ......
   (a) low pressure to high pressure　　(b) high pressure to low pressure
   (c) constant pressure region　　(d) none of these

2. To cool the paramagnetic substance, the process carried out is ......
   (a) isothermal demagnetisation
   (b) adiabatic demagnetisation
   (c) Joule-Thomson throttling
   (d) regenerative cooling

3. Lambda point temperature for liquid helium is about ......
   (a) 10°K　　(b) 4.2°K　　(c) 0.4°K　　(d) 2.19°K

4. For liquid helium II, viscosity ...... with decrease in temperature.
   (a) increases　　　　　　　　(b) does not change
   (c) decreases　　　　　　　　(d) abruptly increases

5. Heat conductivity of liquid helium II is ......
   (a) high　　　　　　　　　　(b) low
   (c) very low　　　　　　　　(d) zero

6. In Joule-Thomson porous plug experiment, all the gases showed cooling effect except ......
   (a) hydrogen　　　　　　　　(b) nitrogen
   (c) oxygen　　　　　　　　　(d) carbon dioxide

ANSWERS

1. (b)	2. (b)	3. (d)	4. (c)	5. (a)	6. (a)		

(B) Short Answer Type Questions :

1. What is Joule-Thomson adiabatic throttling ?
2. What do you mean by regenerative cooling ?
3. State the principle used in Linde's method of liquefaction of air.
4. What is magnetocaloric effect ?
5. What is Lambda point ?
6. State important properties of liquid He II.

(C) Long Answer Type Questions :

1. Describe liquefaction of gas by Joule-Thomson effect.
2. Describe Linde's air liquefier.
3. Explain cooling by adiabatic demagnetisation.
4. Derive expression for fall in temperature due to adiabatic demagnetisation of paramagnetic salt.
5. Describe experimental set-up for adiabatic demagnetisation of paramagnetic substance.
6. Discuss any two properties of liquid He II.

❑❑❑

THERMODYNAMICS

3.1 INTRODUCTION

Thermodynamic system : A definite quantity of matter (solid, liquid or gases) bounded by some closed surface.

Homogeneous system : System is completely uniform throughout.

Heterogeneous system : System consists of two or more phases.

Surrounding : Anything outside the system which can exchange energy with the system.

Thermodynamic variables :

The composition, pressure, volume and temperature are thermodynamic variables or variables of state.

In general, the equation of state is,

$$f(P, V, T) = 0$$

Three classes of system :

(i)   **Open system :** A system can exchange matter and energy with the surrounding.

(ii) **Closed system :** A system which can exchange only energy and not matter with the surrounding.

(iii) **Isolated system :** A thermally isolated system (no exchange of energy and matter).

Thermodynamic equilibrium :

(i)   **Mechanical equilibrium :** No unbalanced forces are acting on the system.

(ii) **Thermal equilibrium :** No temperature difference between the parts of the system.

(iii) **Chemical equilibrium :** No chemical reactions within the system.

Temperature : If a number of systems are in thermal equilibrium, the common property of the system can be represented by a single numerical value called the temperature.

Concept of heat : Heat is defined as the energy in transit. Heat can flow from a body or to a body. It is only used when there is transfer of energy between two or more systems.

3.2 LAWS OF THERMODYNAMICS

3.2.1 Zeroth Law of Thermodynamics

When a hot body A is brought in thermal contact with cold body B, heat flows from A to B and after some time the flow stops. The bodies are then said to be in thermal equilibrium with each other.

Statement : The zeroth law of thermodynamics states that if two bodies A and B are each separately in thermal equilibrium with a third body C, then A and B are also in thermal equilibrium with each other.

Explanation :

Consider two systems A and B are in thermal equilibrium with system C separately as shown in Fig. 3.1 (a). Systems A and B are insulated from each other.

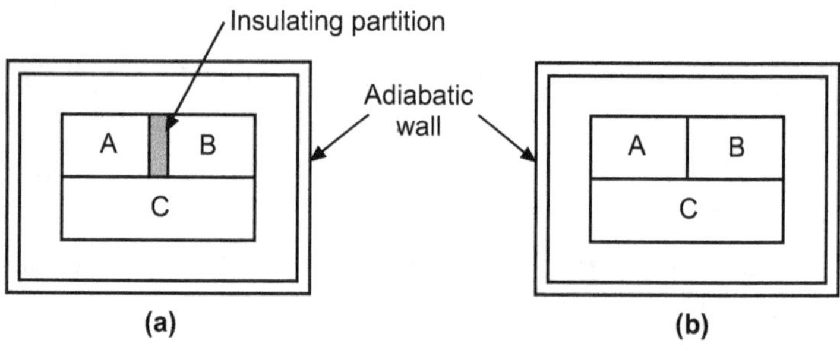

Fig. 3.1

Now if the insulating partition is removed and systems A and B are brought into thermal contact, we find that there is no further change. This means that systems A and B are already in the thermal equilibrium with each other as shown in Fig. 3.1 (b).

Let the temperatures of the systems A, B and C are T_A, T_B and T_C respectively.

As the systems A and C are in thermal equilibrium,

$$T_A = T_C \qquad \text{... (1)}$$

Similarly, as the systems B and C are in thermal equilibrium,

$$T_B = T_C \qquad \text{... (2)}$$

$\therefore$ From equations (1) and (2), we have

$$T_A = T_B$$

This shows that the systems A and B are in thermal equilibrium with one another.

3.2.2 First Law of Thermodynamics

Statement : The amount of heat supplied to a system is equal to the sum of increase in the internal energy of the system and the external work done by the system.

If   dQ – amount of heat absorbed

dU – change in internal energy of the system

dW – external work done, then

$$dQ = dU + dW$$

$\therefore$ $\boxed{dQ = dU + PdV}$ $\qquad$ ... (1)

Significance of the first law :

(i)   It is applicable to any process by which system undergoes a physical or chemical change.

(ii) It introduces the concept of internal energy.

(iii) It depends on the principle of conservation of energy of a system.

3.3 REVERSIBLE AND IRREVERSIBLE PROCESSES

3.3.1 Reversible Process (Change)

The reversible process is the one which can be retraced in the reverse direction exactly along the same path and the system can be brought back to the initial condition by changing the external conditions infinitesimally.

Examples :

(1)　Carnot's cycle.

(2)　Ice melt when a certain amount of heat absorbed by it. The water is converted into ice, if the same amount of heat is removed from it.

(3)　Consider a gas enclosed in a cylinder of perfectly conducting material and immersed in a large tank containing water at constant temperature. If the gas is compressed isothermally (action is performed very slowly) then heat generated will be given to the surrounding water. Again if the gas is allowed to expand isothermally (action performed very slowly) then exactly the same amount of heat will be received during expansion as was given up during compression. Thus all stages of direct process are retraced in the opposite direction and inverse order. Hence infinitesimally slow isothermal expansion and compression of a gas is a reversible process.

The conditions of reversibility are :

(i)　The process should be retraceable in the opposite direction.

(ii)　The heat transferred to the system is equal to the heat transferred by the system.

$$\therefore \qquad dQ = 0$$

$\therefore$ Change in entropy,

$$dS = \frac{dQ}{T} = 0$$

i.e. entropy does not change during reversible process.

(iii)　There should not be dissipative effect like friction, electrical resistance etc. during the process.

(iv)　The reversible process should be performed quasi statically (infinitesimally slow).

3.3.2 Irreversible Process

A process which cannot be retraced in the opposite direction is called irreversible process.

Examples :

1. All natural processes are irreversible because (i) The natural process does not take place quasi-statically and (ii) Dissipative effects, such as friction, viscosity, inelasticity, electric resistance are always present.

2. If the body is rubbed against another, heat is generated due to friction. If the direction of rubbing is reversed, this heat generated will not be absorbed again.

3. If the electrical current is passed through a resistor, heat is generated. Even if the direction of current is reversed, this heat generated will not be absorbed back into resistor.

In the irreversible process :

(i) Heat absorbed $dQ \neq 0$ and is positive.

$$\therefore \quad dS = \frac{dQ}{T} > 0 \text{ i.e. entropy is positive and entropy increases.}$$

(ii) As the heat is supplied from outside, by the surroundings the irreversible process brings about changes in surrounding.

3.4 ISOTHERMAL PROCESS

The process in which the system is perfectly conducting to the surrounding and the temperature of the system remains constant is called isothermal process. In this process, pressure and volume of working substance changes but temperature remains constant.

In isothermal process, following points are noted :

(1) System is perfectly conducting to the surroundings.

i.e. $\quad dQ \neq 0$

(2) The internal energy (dU) of system remains constant.

$$\therefore \quad dQ = dU + dW$$

$$dQ = dW$$

(3) Like P, V and T, there is one more co-ordinate, that is entropy of the system. It is denoted by symbol S.

∴   Change in entropy,

$$dS = \frac{dQ}{T}$$

As $dQ \neq 0, \ dS \neq 0$

∴   Entropy changes during isothermal process.

(4)   The Boyle's law PV = constant is roughly obeyed.

(5)   This is a slow process.

(6)   The isotherms (i.e. graph of P against V and constant T) have small slope i.e. they are less steep.

3.5 ADIABATIC PROCESS

An adiabatic process is a process in which no exchange (i.e. gain or loss) of heat takes place between the system and surrounding.

In adiabatic process, following points are noted :

(1)   The system should be completely insulated from surrounding.

i.e.a $dQ = 0$

(2)   The change in entropy,

$$dS = \frac{dQ}{T} = 0$$

Thus adiabatic process is the process in which entropy always remains constant.

(3)   There is change in temperature (during expansion and compression). Thus Boyle's law is not obeyed.

(4)   Adiabatic process takes place suddenly or quickly.

(5)   The graph plotted between P and V under adiabatic condition is called adiabatic. They are steep.

(6)   As $dQ = 0,$

$$dQ = dU + dW \text{ becomes}$$

$$dU = -dW$$

In adiabatic expansion, work is done by the system and its internal energy decreases while in adiabatic compression work is done on the system and its internal energy increases.

3.6 ADIABATIC RELATIONS

Consider 1 mole of a perfect gas under an adiabatic expansion. It changes from A to B as shown in Fig. 3.2.

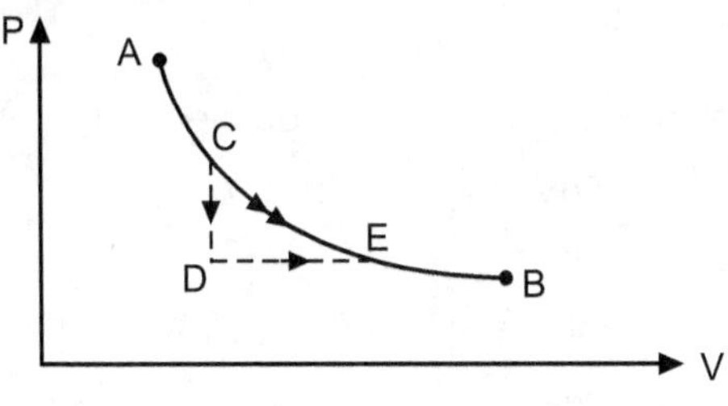

Fig. 3.2

Let CE represent small adiabatic expansion, dV is very small increase in volume and dT is fall in temperature.

The adiabatic change can be done in two steps :

(1)　A fall in temperature $(dT)_V$ at constant volume along CD.

(2)　A fall in temperature $(dT)_P$ at constant pressure along DE.

Let C_P and C_V are specific heat of a gas at constant pressure and volume respectively.

Then :

(a)　Heat absorbed along CD = $C_V\,(dT)_V$.

(b)　Heat absorbed along DE = $C_P\,(dT)_P$.

∴　Total amount of heat absorbed is dQ.

∴ $$dQ = C_V\,(dT)_V + C_P\,(dT)_P$$

But during adiabatic process, no heat gain or loss.

i.e. $$dQ = 0$$

∴ $$0 = C_V\,(dT)_V + C_P\,(dT)_P \qquad \text{... (3.1)}$$

For one mole of gas,

$$PV = RT \qquad \text{... (3.2)}$$

Differentiating at constant volume,

$$V \cdot dP = R\,(dT)_V$$

∴ $$(dT)_V = \frac{V\,dP}{R}$$

Again differentiating equation (3.2) at constant pressure,

$$PdV = R(dT)_P$$

$$\therefore \quad (dT)_P = \frac{PdV}{R}$$

Put the values of $(dT)_V$ and $(dT)_P$ in equation (3.1).

$$\therefore \quad C_V \cdot \frac{VdP}{R} + C_P \frac{PdV}{R} = 0$$

$$\therefore \quad C_V \cdot VdP + C_P \cdot PdV = 0$$

Dividing by V.P. C_V, we get

$$\frac{C_V \cdot V \cdot dP}{V \cdot P \cdot C_V} + \frac{C_P \cdot P \cdot dV}{V \cdot P \cdot C_V} = 0$$

$$\therefore \quad \frac{dP}{P} + \frac{C_P}{C_V}\frac{dV}{V} = 0$$

$$\frac{C_p}{C_V} = \gamma = \text{ratio of specific heats}$$

$$\therefore \quad \frac{dP}{P} + \gamma\frac{dV}{V} = 0$$

Integrating,

$$\log_e P + \gamma \log_e V = \text{constant}$$

$$\therefore \quad \log P + \log V^\gamma = \text{constant}$$

$$\therefore \quad \log PV^\gamma = \text{constant}$$

$$\therefore \quad PV^\gamma = e^{\text{constant}} = \text{constant}$$

$$\therefore \quad \boxed{PV^\gamma = \text{constant}} \qquad \ldots (3.3)$$

This is the relation between P and V.

We know that, $\qquad PV = RT$

$$\therefore \quad P = \frac{RT}{V}$$

Putting in equation (3.3),

$$\left(\frac{RT}{V}\right) \cdot V^\gamma = \text{constant}$$

$$\therefore \quad RT\, V^{\gamma-1} = \text{constant}$$

But γ is constant.

$\therefore$ 　　　　$\boxed{TV^{\gamma-1} = \text{constant}}$ 　　　　... (3.4)

Also 　　　　$V = \dfrac{RT}{P}$ and equation (3.3) becomes

$$P\left[\dfrac{RT}{P}\right]^{\gamma} = \text{constant}$$

$\therefore$ 　　　　$\dfrac{R^{\gamma} \cdot T^{\gamma}}{P^{\gamma-1}} = \text{constant}$

or 　　　　$\dfrac{T^{\gamma}}{P^{\gamma-1}} = \text{constant}$ or $\boxed{\dfrac{P^{\gamma-1}}{T^{\gamma}} = \text{constant}}$ 　　　　... (3.5)

3.7 WORK DONE DURING AN ISOTHERMAL PROCESS

Consider n moles of a perfect gas enclosed in a cylinder of perfectly conducting material. Let the gas expand isothermally at temperature T from its initial state $A(P_1, V_1)$ to final state $B(P_2, V_2)$. When a gas is allowed to expand isothermally, work is done by it.

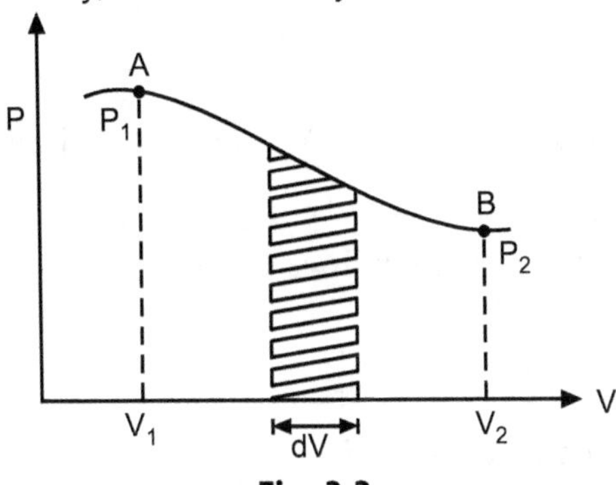

Fig. 3.3

Suppose at some stage, during expansion, the pressure is P and very small change dV takes place in volume. Then the work done by the gas during the change is,

$$dW = PdV$$

$\therefore$ 　　The total work done during the expansion from A to B is,

$$W = \int_{A}^{B} dW = \int_{V_1}^{V_2} PdV$$

But $PV = nRT$, for n moles of a gas.

$$\therefore \quad P = \frac{nRT}{V}$$

$$\therefore \quad W = \int_{V_1}^{V_2} \frac{nRT}{V}\, dV$$

$$= nRT \int_{V_1}^{V_2} \frac{1}{V}\, dV$$

$$= nRT \left[\log_e V\right]_{V_1}^{V_2}$$

$$= nRT \left[\log V_2 - \log V_1\right]$$

$$= nRT \log_e \left(\frac{V_2}{V_1}\right)$$

$$\boxed{W = 2.303\, nRT \log_{10} \left(\frac{V_2}{V_1}\right)}$$

This is an expression for work done W during an isothermal process.

Here, the change in the internal energy of the system is zero (because the temperature remains constant). So the heat absorbed by the system is equal to the work done by it.

3.8 WORK DONE DURING ADIABATIC PROCESS

Consider n moles of a perfect gas enclosed in a cylinder with insulated walls and fitted with a frictionless insulated piston. The gas expand adiabatically from its initial state $(P_1\ V_1\ T_1)$ to final state $(P_2\ V_2\ T_2)$. The change can be supposed to be made up of a number of infinitesimal changes.

Suppose at some stage during the change, pressure is P and a very small increase (change) dV in volume takes place.

Then, work done during this change,

$$dW = PdV$$

∴　The total work done by the gas during the finite change from (P_1V_1) to (P_2V_2) is,

$$W = \int dW = \int_{V_1}^{V_2} P\,dV$$

For an adiabatic process,

$$PV^\gamma = \text{constant} = k$$

∴
$$P = \frac{k}{V^\gamma} = kV^{-\gamma}$$

∴
$$W = \int_{V_1}^{V_2} kV^{-\gamma}\,dV = k\int_{V_1}^{V_2} V^{-\gamma}\,dV$$

$$= k\left[\frac{V^{-\gamma+1}}{-\gamma+1}\right]_{V_1}^{V_2}$$

$$= k\left[\frac{V^{1-\gamma}}{1-\gamma}\right]_{V_1}^{V_2}$$

$$= \frac{1}{1-\gamma}[k\,V_2^{1-\gamma} - k\,V_1^{1-\gamma}] \qquad \ldots (3.6)$$

But
$$PV^\gamma = k$$

∴
$$P_1V_1^\gamma = P_2V_2^\gamma = k$$

∴
$$W = \frac{1}{1-\gamma}[P_2V_2^\gamma \cdot V_2^{1-\gamma} - P_1V_1^\gamma \cdot V_1^{1-\gamma}]$$

∴
$$W = \frac{1}{1-\gamma}[P_2V_2 - P_1V_1] \qquad \ldots (3.7)$$

As for n moles of a gas,

$$PV = nRT$$

∴
$$P_1V_1 = nRT_1 \text{ and } P_2V_2 = nRT_2$$

∴
$$W = \frac{1}{1-\gamma}[nRT_2 - nRT_1]$$

∴
$$\boxed{W = \frac{nR}{1-\gamma}[T_2 - T_1]} \qquad \ldots (3.8)$$

This is the required equation of work done during an adiabatic process.

The nature of curve for adiabatic and isothermal process are as shown in Fig. 3.4.

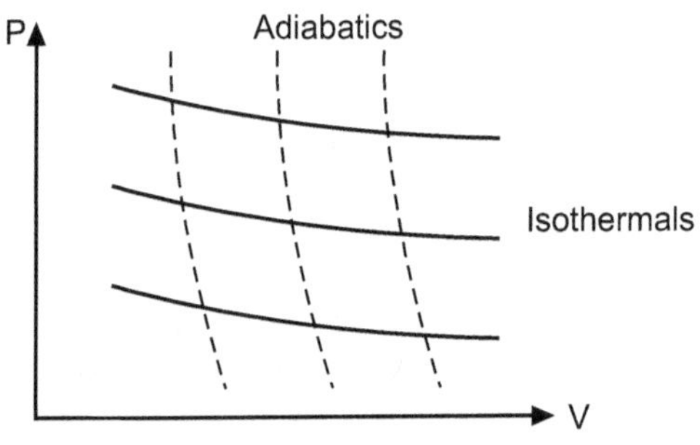

Fig. 3.4

3.9 SECOND LAWS OF THERMODYNAMICS

The first law of thermodynamics is simply the principle of conservation of energy applied to thermodynamical system. But it does not throw light on conditions, under which body absorbs heat. It does not give the direction in which heat transfer takes place.

The second law of thermodynamics gives the direction in which the transfer of heat takes place.

A number of scientists like Kelvin, Planck and Clausius stated the law. They all are equivalent and are known as second laws of thermodynamics.

(a) **Kelvin statement :** It is impossible to get a continuous supply of work from a body by cooling it to a temperature lower than that of its surrounding.

(b) **Kelvin-Planck statement :** It is impossible to get a continuous supply of work from a body which can transfer heat with a single heat reservoir.

(c) **Clausius statement :** It is impossible for a self acting machine working in a cyclic process, unaided by external agency, to transfer heat from a body at a lower temperature to a body at a higher temperature.

3.10 THIRD LAW OF THERMODYNAMICS

The heat capacities of all solids tend to zero as absolute zero of temperature is approached and that the internal energies and entropies of all substances become equal there, approaching their common value asymptotically tending to zero.

3.11 ENTROPY CHANGE IN REVERSIBLE PROCESS

Consider a complete reversible Carnot's cycle ABCD as shown in Fig. 3.5. Here two isothermals are AB and CD at temperatures T_1 and T_2 respectively and two adiabatics BC and DA.

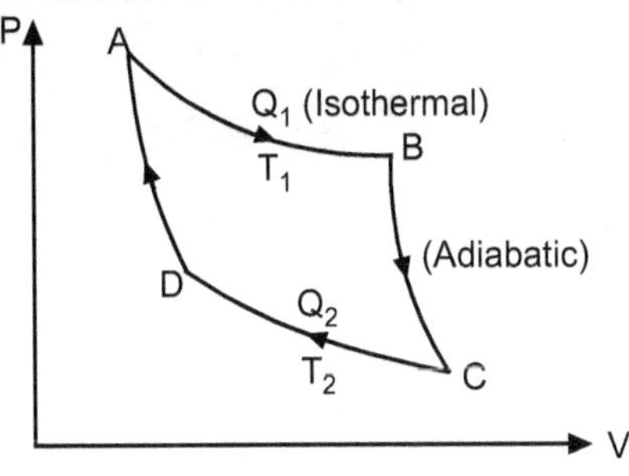

Fig. 3.5

(i) **Isothermal expansion AB :** Let Q_1 be the amount of heat absorbed by the working substance in going from state A to state B during isothermal expansion AB at constant temperature T_1. The increase in entropy of working substance is,

$$\int_A^B dS = +\frac{Q_1}{T_1} \qquad \ldots (3.9)$$

(ii) **Adiabatic expansion BC :** In going from state B to state C along adiabatic BC, there is no change in entropy of working substance, but temperature falls from T_1 to T_2 due to expansion.

$$\int_B^C dS = 0 \qquad \ldots (3.10)$$

(iii) **Isothermal compression CD :** In going from C to D along isothermal CD, the working substance rejects heat Q_2 to the sink at temperature T_2. The entropy of the working substance decreases and the change in entropy is given by

$$\int_C^D dS = -\frac{Q_2}{T_2} \qquad \ldots (3.11)$$

(iv) Adiabatic compression DA : In going from D to A along the adiabatic DA, there is no change in entropy but temperature rises from T_2 to T_1.

$$\int_D^A dS = 0 \qquad \ldots (3.12)$$

$\therefore$ The net gain in entropy in the whole cycle ABCDA,

$$\oint dS = \int_A^B dS + \int_B^C dS + \int_C^D dS + \int_D^A dS$$

$\therefore$

$$\oint dS = \frac{Q_1}{T_1} - \frac{Q_2}{T_2} \text{ (from 3.9, 3.10, 3.11, 3.12) } \ldots (3.13)$$

But for reversible Carnot's cycle,

$$\frac{Q_1}{T_1} = \frac{Q_2}{T_2}$$

$\therefore$

$$\oint dS = 0$$

Thus in a reversible process, the entropy of the system remains constant or unchanged or total change in entropy is always zero.

3.12 ENTROPY CHANGE IN IRREVERSIBLE PROCESS

Consider the natural process of conduction of heat from a body A at temperature T_1 to another body B at lower temperature T_2. This process is irreversible.

Since heat always flows from higher to a lower temperature the quantity of heat thus transmitted be Q, then

Decrease in entropy of a hotter body $= \dfrac{Q}{T_1}$

and Increase in entropy of a colder body $= \dfrac{Q}{T_2}$

$\therefore$ Net increase in entropy of the system,

$$dS = \frac{Q}{T_2} - \frac{Q}{T_1}$$

$$= +\text{ve quantity} \qquad (\because T_1 > T_2)$$

Thus the entropy of system increases in all irreversible processes. This is known as law of principle of increase of entropy.

All natural processes taking place in the universe are irreversible. It means that entropy of universe increases.

$$dS \text{ [universe]} \geq 0$$

SOLVED PROBLEMS

Problem 3.1 : *Calculate the work done when kilomole of a perfect gas expands isothermally at 25 °C to double its original value.*

$[R = 8.3 \text{ J/kmole } °K]$

Solution :

$$R = 8.3 \text{ J/kmol°K}$$
$$T_1 = 25°C = 298°K$$
$$V_2 = 2V_1$$
$$\therefore \quad \frac{V_2}{V_1} = 2, \quad n = 1$$
$$W = 2.303 \, nRT \log_{10} \frac{V_2}{V_1}$$
$$= 2.303 \times 1 \times 8.3 \times 298 \times 0.3010$$
$$W = \mathbf{1.714 \times 10^3 \ J}$$

Problem 3.2 : *The temperature of 10 gm of air is raised by 1 °C at constant volume. Calculate increase in its internal energy. (1 cal = 4.18 J)*

Given : $C_v = 0.172$ cal/gm °C and 1 cal = 4.18 J

Solution :

$$dQ = dU + dW$$
$$dU = dQ - dW$$
$$= dQ - PdV$$

The air is heated at constant volume.

Hence, $\quad dV = 0$

$\therefore \quad dU = dQ$

$$= \text{Mass of air} \times C_v \times \text{Change in temperature}$$
$$= 10 \times 0.172 \times 1$$
$$= 1.72 \text{ cal}$$
$$= 1.72 \times 4.18$$
$$= \mathbf{7.19 \ J}$$

Problem 3.3 : *Air at N.T.P. is compressed adiabatically to half of its volume. Calculate the change in its temperature.*

Solution : Let, the initial temperature be T_1 °K and final temperature T_2 °K.

$$\text{Initial volume} = V_1$$

$$\therefore \quad \text{Final volume} = V_2 = \frac{V_1}{2}$$

During an adiabatic process,

$$T_1 V_1^{\gamma-1} = T_2 V_2^{\gamma-1}$$

$$T_2 = T_1 \left[\frac{V_1}{V_2}\right]^{\gamma-1}$$

$$\therefore \quad T_2 = T_1 \, [2]^{\gamma-1}$$

But γ for air $= 1.4$

$$\therefore \quad T_2 = T_1 \, [2]^{0.4}$$

$$= 1.319 \, T_1$$

$\therefore$ Change in temperature

$$= T_2 - T_1$$

$$= 1.319 \, T_1 - T_1$$

$$= \mathbf{0.319 \, T_1 \, K}$$

Problem 3.4 : *A quantity of air at 27°C and atmospheric pressure is suddenly compressed to half of its original volume.*

Find the final (i) pressure, (ii) temperature.

(**Given :** $\gamma = 1.4$, $2^{1.4} = 2.64$)

Solution : (i) $P_1 = 1$ atmosphere, $P_2 = ?$, $\gamma = 1.4$

$$V_1 = V \quad \therefore \ V_2 = \frac{V}{2}$$

During sudden compression, the process is adiabatic.

$$\therefore \quad P_1 V_1^{\gamma} = P_2 V_2^{\gamma}$$

$$\therefore \quad P_2 = P_1 \left[\frac{V_1}{V_2}\right]^{\gamma}$$

$$= 1 \, [2]^{1.4}$$

$$P_2 = \textbf{2.64 atmospheres}$$

(ii) $V_1 = V, \quad V_2 = \dfrac{V}{2}, \quad T_1 = 300 \text{ K}, \quad T_2 = ?, \quad \gamma = 1.4.$

$$T_1 V_1^{\gamma-1} = T_2 V_2^{\gamma-1}$$

$$T_2 = T_1 \left(\frac{V_1}{V_2}\right)^{\gamma-1}$$

$$= T_1 \, [2]^{1.4-1}$$

$$= 300 \, [2]^{0.4}$$

$$= 395.9 \text{ K}$$

$$T_2 = \textbf{122.9°C}$$

Problem 3.5 : *A certain mass of gas at NTP is expanded to three times its volume under adiabatic conditions. Calculate the resulting temperature and pressure. γ for the gas is 1.40.*

Solution : (a) $\quad V_1 = V, \; V_2 = 3V, \; T_1 = 273 \text{ K}, \; T_2 = ?$

$$T_1 V_1^{\gamma-1} = T_2 V_2^{\gamma-1}$$

$$\therefore \quad T_2 = T_1 \left[\frac{V_1}{V_2}\right]^{\gamma-1} = 273 \left[\frac{1}{3}\right]^{1.4-1}$$

$$T_2 = 176 \text{ K} = -\textbf{97°C}$$

(b) $\quad V_1 = V, \; V_2 = 3V, \; P_1 = 1 \text{ atm}, \; P_2 = ?$

$$P_1 V_1^{\gamma} = P_2 V_2^{\gamma}$$

$$\therefore \quad P_2 = P_1 \left[\frac{V_1}{V_2}\right]^{\gamma}$$

$$\therefore \quad P_2 = 1 \left(\frac{1}{3}\right)^{1.4} = \textbf{0.2148 atm.}$$

Problem 3.6 : *A gas occupying 1 litre at 80 cm of Hg pressure is expanded adiabatically to 1190 c.c. If the pressure falls at 60 cm of Hg in the process, deduce the value of γ.*

Given : P_1 = 80 cm of Hg, P_2 = 60 cm of Hg, V_1 = 1000 c.c., V_2 = 1190 c.c.

Solution :
$$P_1 V_1^{\gamma} = P_2 V_2^{\gamma}$$

$$\left(\frac{P_1}{P_2}\right) = \left(\frac{V_2}{V_1}\right)^{\gamma}$$

Taking logarithms,

$$\log\left(\frac{P_1}{P_2}\right) = \gamma \log\left(\frac{V_2}{V_1}\right)$$

$\therefore$
$$\gamma = \frac{\log\left(\dfrac{P_1}{P_2}\right)}{\log\left(\dfrac{V_2}{V_1}\right)}$$

$$= \frac{\log\left(\dfrac{80}{60}\right)}{\log\left(\dfrac{1190}{1000}\right)}$$

$$= \frac{0.1249}{0.0755}$$

$$\gamma = \mathbf{1.787}$$

Problem 3.7 : *Calculate the rise in temperature of a gas initially at 27 °C, if its pressure is suddenly doubled, $\gamma = 1.4$.*

Solution :
$$\frac{P_1^{\gamma-1}}{T_1^{\gamma}} = \frac{P_2^{\gamma-1}}{T_2^{\gamma}}$$

$$\left(\frac{P_2}{P_1}\right)^{\gamma-1} = \left(\frac{T_2}{T_1}\right)^{\gamma}$$

$P_2 = 2P_1$, $T_1 = 27°C = 273 + 27 = 300$ K, $T_2 = ?$

$$\left(\frac{2P_1}{P_1}\right)^{1.4-1} = \left(\frac{T_2}{300}\right)^{1.4}$$

$$(2)^{0.4} = \left(\frac{T_2}{300}\right)^{1.4}$$

$$0.4 \log 2 = 1.4 \left[\log T_2 - \log 300\right]$$

$$0.4 \times 0.3010 = 1.4 \left[\log T_2 - 2.4771\right]$$

$$\log T_2 = \frac{0.4 \times 0.3010}{1.4} + 2.4771$$

$$= 0.086 + 2.4771$$

$$= \mathbf{2.5631}$$

$$T_2 = 365.7 \text{ K}$$

$$= 365.7 - 273 = \mathbf{92.7°C}$$

$$\therefore \quad \text{Rise in temperature} = (92.7 - 27)°C$$

$$= \mathbf{65.7°C}$$

EXERCISES

(A) Multiple Choice Questions :

1. Out of the following, the physical quantity that relates with first law of thermodynamics is ......

   (a) temperature　　　　　　(b) pressure

   (c) energy　　　　　　　　(d) number of moles

2. All natural processes are ......

   (a) reversible　　　　　　(b) irreversible

   (c) isothermal　　　　　　(d) isobaric

3. Entropy of reversible process ......

   (a) increases　　　　　　(b) decreases

   (c) remains constant　　　(d) zero

4. The adiabatic relation between temperature and volume is ......

   (a) $TV^{\gamma} = $ constant　　　　(b) $TV^{\gamma+1} = $ constant

   (c) $TV^{\gamma-1} = $ constant　　　(d) $TV^{1-\gamma} = $ constant

5. In an isothermal process ...... remains constant.

   (a) pressure　　　　　　　(b) temperature

   (c) volume　　　　　　　　(d) pressure and temperature

6. According to second law of thermodynamics, heat by itself cannot pass from ......

(a) colder to hotter (b) hotter to colder

(c) lighter to heavier (d) heavier to lighter

7. The zeroth law of thermodynamics leads to definition of the term ......

(a) pressure (b) volume

(c) entropy (d) temperature

8. First law of thermodynamics introduces the concept of ......

(a) temperature (b) pressure

(c) internal energy (d) entropy

9. A process which cannot be retraced in the opposite direction is called ......

(a) reversible process (b) irreversible process

(c) isochoric process (d) none of these

10. In an adiabatic expansion, internal energy ......

(a) increases (b) decreases

(c) remains constant (d) becomes equal to zero

11. In an isothermal process, internal energy ......

(a) increases (b) decreases

(c) remains constant (d) none of these

12. According to third law of thermodynamics, the heat capacities of all solids tend to ...... as absolute zero of temperature is approached.

(a) zero (b) unity

(c) infinite (d) integer

ANSWERS

1. (c)	2. (b)	3. (c)	4. (c)	5. (b)	6. (a)	7. (d)	8. (c)
9. (b)	10. (b)	11. (c)	12. (a)				

(B) Short Answer Type Questions :

1. State and explain zeroth law of thermodynamics.

2. State and give significance of first law of thermodynamics.

3. What is reversible process ? State the conditions of reversibility.

4. What is irreversible process ? Explain using two examples.

5. Write a note on isothermal process.

6. Write a note on adiabatic process.

7. Why entropy of irreversible process increases ? Explain.

(C) Long Answer Type Questions :

1. Write a note on isothermal process and derive an expression for the work done in isothermal process.

2. Write a note on adiabatic process and derive an expression for the work done in adiabatic process.

3. Show that PV^{γ} = constant for adiabatic process.

4. Show that energy change in reversible process is equal to zero.

(D) Problems for Practice :

1. 1 kg of water is boiled under a pressure of 12 atm at 120°C. If the volumes occupied by water and steam under given conditions are respectively 10^{-3} m^3 and 0.842 m^3, calculate (a) the work done, (b) increase in internal energy.   (**Ans.** 1.70×10^5 J, 2.03×10^6 J)

2. A certain mass of gas at NTP is expanded to three times its volume under adiabatic conditions. Calculate the resulting temperature and pressure.   (**Ans.** 0.215 atmosphere, $-$ 97°C)

3. A quantity of dry air at 27°C is compressed : (i) slowly, (ii) suddenly to $\dfrac{1^{rd}}{3}$ of its volume. Find the change in temperature in each case $(\gamma = 1.4)$.

   (Ans. (i) No change in temperature, (ii) 165.5 K)

4. A certain mass of gas at NTP is expanded to three times its volume under adiabatic conditions. Calculate the resulting temperature and pressure $(\gamma = 1.4)$.　　　　**(Ans.** – 97°C, 0.2148 atmosphere)

4

CHAPTER

HEAT ENGINES

- Any device which converts heat into mechanical work is called a heat engine.

- Heat engines in their operation absorb heat at higher temperature, convert part of it into mechanical work and reject the remaining heat at a lower temperature. In this process, a working substance is used. In steam engine, the working substance is water vapour and in all gas engines, the working substance is combustible mixture of gases.

- In any heat engine, the working substance undergoes certain changes of pressure, volume and temperature and then returns to the initial state. The complete change in state of working substance from its initial state and back to its starting state is called one cycle of operation.

- The efficiency of heat engine is given as,

$$\eta = \frac{Q_1 - Q_2}{Q_1}$$

where Q_1 – heat absorbed from source at higher temperature

   Q_2 – heat rejected to a sink at lower temperature.

Since $(Q_1 - Q_2) < Q_1$, the efficiency can never be 100%.

4.2 CARNOT'S IDEAL HEAT ENGINE

- In 1824, Sadi Carnot conceived a theoretical engine which has maximum efficiency and it is an ideal engine.

- Following are the parts of Carnot's engine :

   1. **Cylinder :** A cylinder having perfectly non-conducting walls, a perfectly conducting base and is provided with a perfectly non-conducting piston which moves without friction in the cylinder.

(4.1)

The cylinder contains one mole of perfect gas as the working substance.

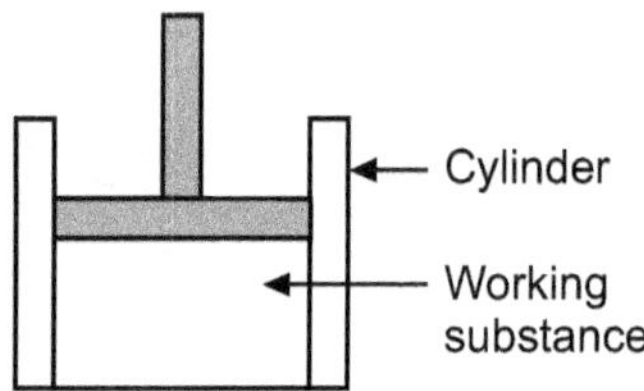

Fig. 4.1 (a)

2. **Source :** A reservoir maintained at a constant higher temperature T_1 from which the engine can draw heat by perfect conduction. It has infinite thermal capacity and any amount of heat can be drawn from it.

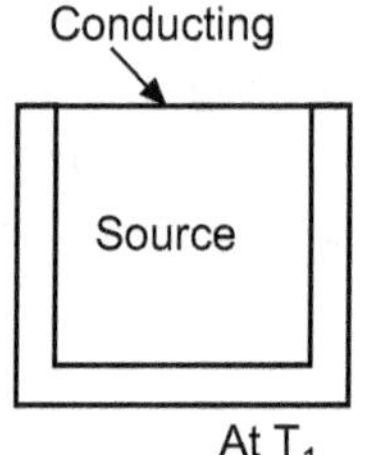

Fig. 4.1 (b)

3. **Heat insulating stand :** A perfectly non-conducting platform acts as a stand for adiabatic processes.

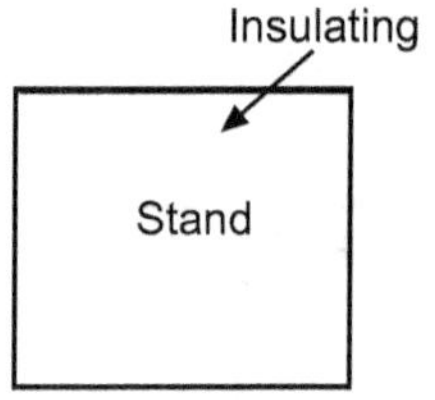

Fig. 4.1 (c)

4. **Sink :** A reservoir maintained at constant lower temperature T_2 $(T_2 < T_1)$ to which the heat engine can reject any amount of heat. The heat capacity of sink is infinite, so that its temperature remains constant at T_2, no matter how much heat is given to it.

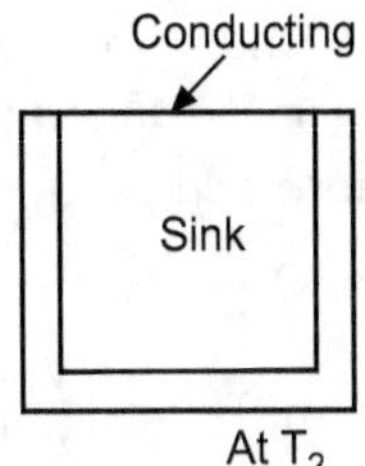

Fig. 4.1 (d)

4.2.1 Working of Carnot's Cycle

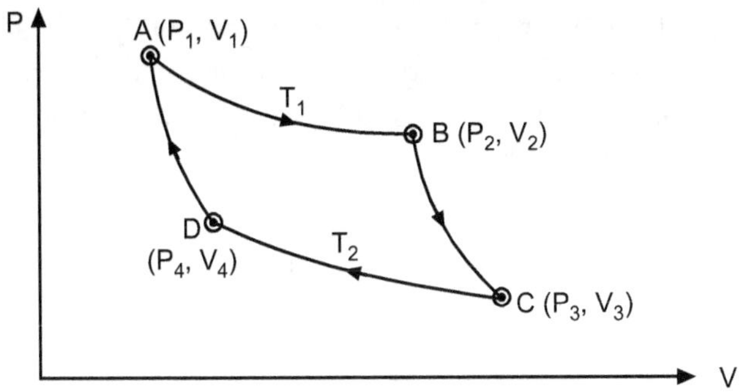

Fig. 4.2

The working of Carnot's heat engine can be represented on the above indicator diagram as Carnot's cycle.

A perfect gas enclosed in a cylinder is taken through four steps of operations. These four steps make one complete cycle known as Carnot's cycle.

1. Isothermal expansion :

The cylinder is kept on the source. The cylinder contains gas having the pressure P_1 and volume V_1.

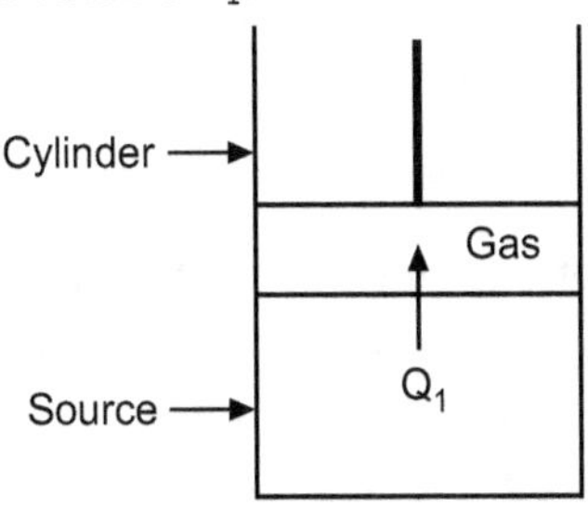

Fig. 4.3

The gas is allowed to expand at constant temperature T_1. The final pressure is P_2 and volume is V_2. During this expansion the work done W_1 is equal to the heat absorbed i.e. Q_1.

$$\therefore \qquad W_1 = Q_1$$

$$= nRT_1 \, \log_e \left(\frac{V_2}{V_1}\right) \qquad \ldots (4.1)$$

This is shown by AB in Fig. 4.2.

2. Adiabatic expansion :

The cylinder is kept on an insulator stand and gas is allowed to expand adiabatically. During this expansion, the initial value is $B(P_2, V_2)$ and the final value is $C(P_3, V_3)$, the temperature of the gas drops down from T_1 to T_2.

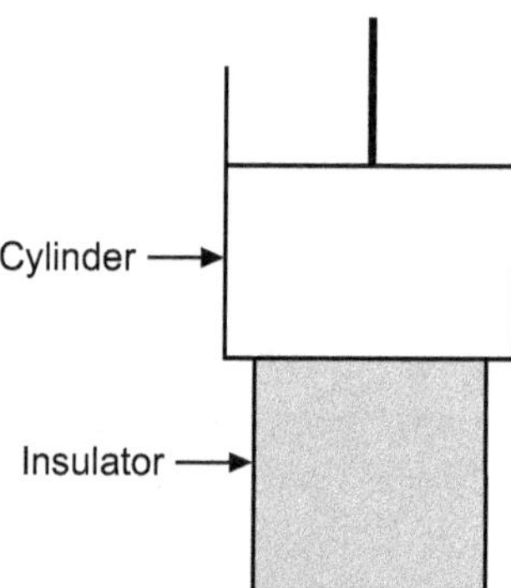

Fig. 4.4

The work done W_2 during this expansion is,

$$W_2 = \frac{R}{\gamma - 1} (T_1 - T_2) \qquad \ldots (4.2)$$

This is shown by BC in Fig. 4.2.

3. Isothermal compression :

The cylinder is kept on the sink and the gas is slowly compressed by moving the piston downward then pressure and volume change from $C(P_3, V_3)$ to $D(P_4, V_4)$. During which the amount of heat Q_2 is given out at constant temperature T_2. The amount of heat given out is equal to the work done W_3 on the gas.

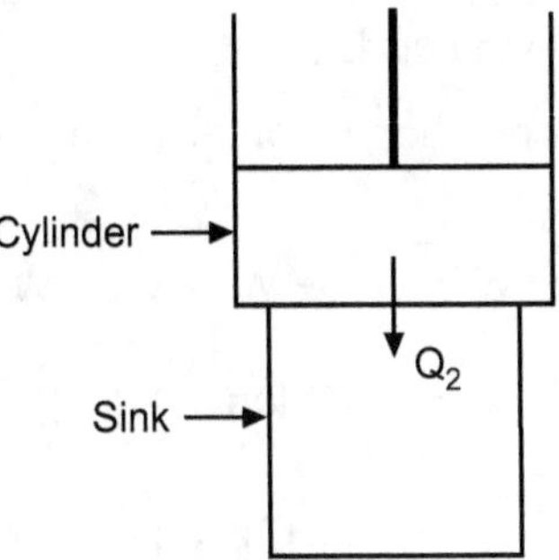

Fig. 4.5

$$\therefore \qquad W_3 = Q_2$$

$$\therefore \qquad W_3 = nRT_2 \log_e\left(\frac{V_4}{V_3}\right)$$

$$\therefore \qquad W_3 = -nRT_2 \log_e\left(\frac{V_3}{V_4}\right) \qquad (\because V_3 > V_4) \ \dots (4.3)$$

This is shown by CD in Fig. 4.2.

4. Adiabatic compression :

The cylinder is placed on an insulator and the gas is compressed to attain the initial state $A(P_1, V_1)$ at temperature T_1.

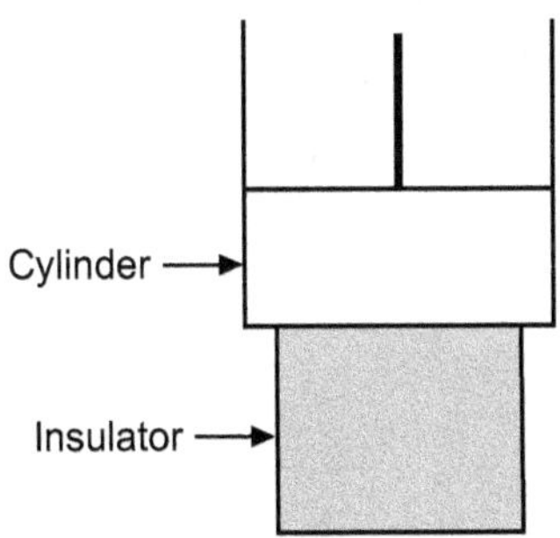

Fig. 4.6

The work done W_4 on the gas is,

$$W_4 = \frac{R}{\gamma - 1}(T_2 - T_1)$$

$$\therefore \qquad W_4 = \frac{-R}{\gamma - 1}(T_1 - T_2) \qquad \dots (4.4)$$

This is shown by DA in Fig. 4.2.

Thus one cycle of operation is completed and during this process, the total work done is,

$$W = W_1 + W_2 + W_3 + W_4$$

$$= nRT_1 \log_e\left(\frac{V_2}{V_1}\right) + \frac{R}{\gamma - 1}(T_1 - T_2)$$

$$- nRT_2 \log_e\left(\frac{V_3}{V_4}\right) - \frac{R}{\gamma - 1}(T_1 - T_2)$$

$$\therefore \quad W = nRT_1 \log_e\left(\frac{V_2}{V_1}\right) - nRT_2 \log_e\left(\frac{V_3}{V_4}\right)$$

From adiabatic relations, we can get

$$\left(\frac{V_2}{V_1}\right) = \left(\frac{V_3}{V_4}\right)$$

$$\therefore \quad \boxed{W = nR \log_e\left(\frac{V_2}{V_1}\right)(T_1 - T_2)} \qquad \text{... (4.5)}$$

We know that, $\quad W = Q_1 - Q_2 \qquad$ ... (4.6)

This gives work done W in one complete cycle.

The efficiency η is given as,

$$\eta = \frac{\text{Work done}}{\text{Heat absorbed}} = \frac{Q_1 - Q_2}{Q_1}$$

$$= \frac{W}{Q_1} = \frac{nR \log_e\left(\frac{V_2}{V_1}\right)(T_1 - T_2)}{nRT_1 \log_e\left(\frac{V_2}{V_1}\right)}$$

$$\eta = \frac{T_1 - T_2}{T_1}$$

$$\therefore \quad \boxed{\eta = 1 - \frac{T_2}{T_1}} \qquad \text{... (4.7)}$$

The expression for efficiency η shows that :

(a) Efficiency η is less than 1 and hence an ideal engine is not 100% efficient.

(b) Efficiency of the engine depends upon two temperatures namely the temperature of source and the sink.

(c)　Efficiency is independent on the nature of the working substance.

Work done during one complete cycle is,

$$W = \text{Area ABCDA}$$

4.3 OTTO CYCLE AND ITS EFFICIENCY

In the Otto engine, air is the working substance and petrol vapour acts as the fuel. Following conditions are assumed in Otto cycle :

(i)　The working substance is air all the time and it behaves as a perfect gas.

(ii)　There is no friction.

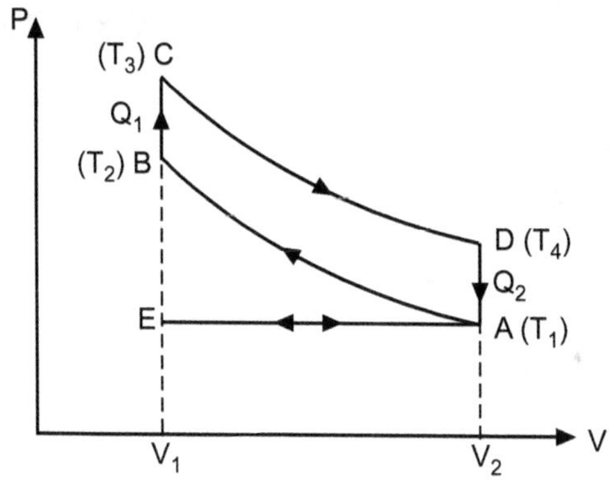

Fig. 4.7

On the basis of these assumptions, Otto cycle is divided into following operations :

1.　E to A represent the charging stroke. The mixture of air and petrol vapour is allowed to enter into the cylinder at atmospheric pressure. The final volume is V_2 at temperature T_1.

2.　A to B represents the adiabatic compression. There is no friction and no flow of heat through the walls of the cylinder. The volume changes from V_2 to V_1 and the temperature changes from T_1 to T_2. There is a change in pressure also. This represents the compression stroke. The compression ratio is nearly 8 i.e. $\dfrac{V_2}{V_1} = 8$.

The pressure changes from one atmosphere to 18 atmosphere.

3. B to C represents the ignition stage. A spark is produced and the mixture of air and petrol vapours is ignited. The pressure increases from 18 to about 80 atmospheres and the temperature changes from T_2 to T_3 (from 500°C to 2000°C).

4. C to D represents the working stroke. The gas expands adiabatically and the engine works. The volume changes from V_1 to V_2. The pressure and temperature decrease.

5. As the point D is reached, the exhaust valve opens. D to A represents the change of pressure to the atmospheric pressure and the temperature changes from T_4 to T_1.

6. A to E represents the exhaust stroke. The exhaust gases are completely discharged from the cylinder. Thus, the initial condition of the engine is restored.

4.3.1 Efficiency of Otto Engine

Consider 1 gram of the working substance is used in the process. A quantity of heat Q_1 is absorbed at higher temperature from B to C.

$$Q_1 = 1 \times C_v (T_3 - T_2) \qquad \ldots (4.8)$$

A quantity of heat Q_2 is rejected at lower temperature from D to A.

$$Q_2 = 1 \times C_v (T_4 - T_1) \qquad \ldots (4.9)$$

$$\therefore \qquad \frac{Q_2}{Q_1} = \frac{T_4 - T_1}{T_3 - T_2} \qquad \ldots (4.10)$$

$$\therefore \quad \text{Efficiency,} \qquad \eta = 1 - \frac{Q_2}{Q_1} = 1 - \frac{T_4 - T_1}{T_3 - T_2} \qquad \ldots (4.11)$$

The points D and C lie on the same adiabatic.

$$\therefore \qquad T_4 V_2^{\gamma-1} = T_3 V_1^{\gamma-1} \qquad \ldots (4.12)$$

Also, the points A and B lie on the same adiabatic,

$$T_1 V_2^{\gamma-1} = T_2 V_1^{\gamma-1} \qquad \ldots (4.13)$$

Subtracting (4.13) from (4.12), we get

$$(T_4 - T_1) V_2^{\gamma-1} = (T_3 - T_2) V_1^{\gamma-1}$$

$$\therefore \qquad \frac{(T_4 - T_1)}{(T_3 - T_2)} = \left(\frac{V_1}{V_2}\right)^{\gamma-1} \qquad \ldots (4.14)$$

∴　Equation (4.11) becomes,

$$\eta = 1 - \left(\frac{V_1}{V_2}\right)^{\gamma-1}$$

$$= 1 - \left(\frac{1}{V_2/V_1}\right)^{\gamma-1}$$

$$\boxed{\eta = 1 - \left(\frac{1}{\rho}\right)^{\gamma-1}}$$
　　　　　　　　　... (4.15)

where $\dfrac{V_2}{V_1} = \rho$, is the adiabatic compression ratio.

In the actual petrol engine, ρ cannot be made greater than 10. If ρ is more than 10, the mixture ignites itself due to compression.

∴　If (i)　　　　　$\rho = 6$ and $\gamma = 1.4$

$$\eta = 1 - \left(\frac{1}{6}\right)^{1.4-1}$$

$$= 0.5116 = 51.16\%$$

(ii)　　　　　$\rho = 9$ and $\gamma = 1.4$

$$\eta = 1 - \left(\frac{1}{9}\right)^{1.4-1}$$

$$= 0.5847 = 58.47\%$$

4.4 DIESEL CYCLE AND ITS EFFICIENCY

In case of diesel engine, air is admitted into the cylinder in the beginning. This air is compressed adiabatically so that the temperature is high enough to ignite the oil sprayed into the cylinder. The ideal conditions are assumed as :

(i)　The working substance is air all the time and it behaves as a perfect gas.

(ii)　There is no friction. The cycle is called diesel cycle.

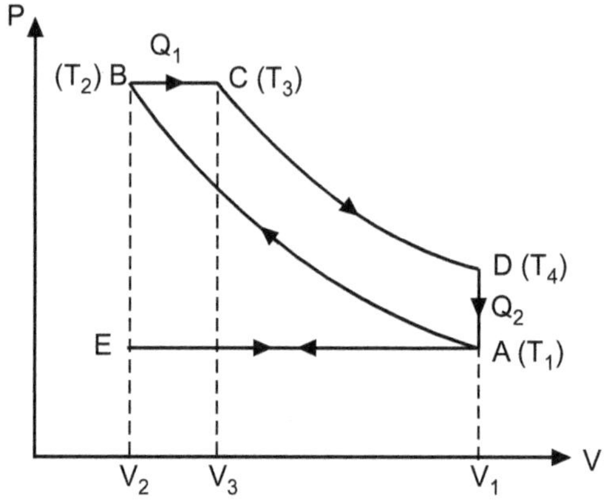

Fig. 4.8

1. E to A represents the intake of air and its volume changes to V_1. This is the charging stroke.

2. A to B represents the compression stroke. Air is compressed adiabatically. The temperature changes from T_1 to T_2 and the volume changes from V_1 to V_2. The temperature rises to about 1000°C and the pressure to about 40 atmospheres.

3. B to C represents the stage when the oil is sprayed into the cylinder. The oil burns immediately. The pressure is maintained constant. The temperature changes from T_2 to T_3 (to 2000°C) and the volume changes from V_2 to V_3.

4. C to D represents the working stroke. The mixture of air and diesel oil vapour expand adiabatically.

5. As the point D is reached, the exhaust valve opens and the pressure drops to the point A. The volume remains constant, but temperature and pressure decrease.

6. A to E represents the exhaust stroke. The unburnt vapours of the oil and the mixture of gases in the cylinder are exhausted out of the cylinder.

Efficiency of the Diesel engine :

From B to C, the pressure remains constant. Considering 1 gram of the working substance, the quantity of heat absorbed,

$$Q_1 = 1 \times C_p (T_3 - T_2)$$

From D to A, the volume remains constant.

The quantity of heat rejected,

$$Q_2 = 1 \times C_v (T_4 - T_1)$$

∴　　　Efficiency, $\eta = 1 - \dfrac{Q_2}{Q_1}$

$$= 1 - \dfrac{C_v (T_4 - T_1)}{C_p (T_3 - T_2)}$$

$$= 1 - \dfrac{1}{\gamma}\left[\dfrac{T_4 - T_1}{T_3 - T_2}\right] \qquad \text{... (4.16)}$$

Let　　　　　　$\rho = \dfrac{V_1}{V_2}$ is adiabatic expansion ratio and

$$e = \dfrac{V_3}{V_2} \text{ is combustion expansion ratio or fuel}$$

cut-off ratio

To evaluate $\left(\dfrac{T_4 - T_1}{T_3 - T_2}\right)$, all temperatures are to be expressed in terms

of T_2.

1.　The points A and B are on the same adiabatic,

$$T_1 V_1^{\gamma-1} = T_2 V_2^{\gamma-1}$$

∴　　　　　$T_1 = T_2\left[\dfrac{V_2}{V_1}\right]^{\gamma-1}$

$$= T_2\left[\dfrac{1}{\rho}\right]^{\gamma-1} \qquad \text{... (4.17)}$$

2.　The points B and C are at the same pressure,

$$\dfrac{T_3}{V_3} = \dfrac{T_2}{V_2}$$

∴　　　　　$T_3 = T_2\left[\dfrac{V_3}{V_2}\right] = T_2 [e] \qquad \text{... (4.18)}$

3.　The points C and D are on the same adiabatic.

∴　　　　　$T_4 V_4^{\gamma-1} = T_3 V_3^{\gamma-1}$

But　　　　　$V_4 = V_1$

$$\therefore \quad T_4 = T_3 \left[\frac{V_3}{V_1}\right]^{\gamma-1}$$

$$= T_3 \left[\frac{V_3}{V_2} \times \frac{V_2}{V_1}\right]^{\gamma-1}$$

$$= T_3 \left[\frac{V_3}{V_2}\right]^{\gamma-1} \left[\frac{V_2}{V_1}\right]^{\gamma-1}$$

$$= T_3 \, [e]^{\gamma-1} \left[\frac{1}{\rho}\right]^{\gamma-1}$$

$$= T_2 \, [e] \, [e]^{\gamma-1} \left[\frac{1}{\rho}\right]^{\gamma-1} \quad \ldots \text{(from equation 4.18)}$$

$$= T_2 \, [e]^{\gamma} \left[\frac{1}{\rho}\right]^{\gamma-1} \qquad \ldots (4.19)$$

$\therefore$ Equation (4.16) becomes

$$\eta = 1 - \frac{1}{\gamma}\left[\frac{T_4 - T_1}{T_3 - T_2}\right]$$

$$= 1 - \frac{1}{\gamma}\left[\frac{T_2 \, [e]^{\gamma} \left[\frac{1}{\rho}\right]^{\gamma-1} - T_2 \left[\frac{1}{\rho}\right]^{\gamma-1}}{T_2 \, [e] - T_2}\right]$$

$$= 1 - \frac{1}{\gamma}\left(\frac{1}{\rho}\right)^{\gamma-1}\left[\frac{T_2 \, [e]^{\gamma} - T_2}{T_2 \, [e] - T_2}\right]$$

$$= 1 - \frac{1}{\gamma}\left(\frac{1}{\rho}\right)^{\gamma-1}\left[\frac{T_2 \, [e^{\gamma} - 1]}{T_2 \, [e - 1]}\right]$$

$$\therefore \quad \eta = 1 - \frac{1}{\gamma}\left(\frac{1}{\rho}\right)^{\gamma-1}\left[\frac{e^{\gamma} - 1}{e - 1}\right] \qquad \ldots(4.20)$$

For the same compression ratio, the efficiency of an Otto engine is more than Diesel engine.

In practice the compression ratio for an Otto engine is from 7 to 9 and for Diesel engine it is from 15 to 20.

Due to higher compression ratio, an actual Diesel engine has higher efficiency than the Otto (petrol) engine. The cylinder must be strong enough to withstand with high pressure.

4.5 COMPARISON BETWEEN OTTO AND DIESEL ENGINES

Otto engine	Diesel engine
1. In Otto (petrol) engine the mixture of air and petrol vapour is fed into the cylinder.	1. In Diesel engine, the oil (diesel) is sprayed onto the compressed air inside the cylinder.
2. It is an ignition engine. The combustion is initiated using a spark plug.	2. It is a compression engine. Auto ignition takes place due to compression of air.
3. Heat transfers at constant volume.	3. Heat transfers at constant pressure.
4. Compression ratio should not be higher than 10.	4. Compression ratio is in the range 15-20.
5. The efficiency is low as compared to the Diesel engine (due to low compression ratio).	5. The efficiency is high as compared to the Otto engine (due to high compression ratio).
6. Otto engines are lighter in weight.	6. Diesel engines are heavier in weight.
7. Used in comparatively lighter vehicles like cars, scooters etc.	7. Mostly used in heavy vehicles like trucks, buses etc.

SOLVED PROBLEMS

Problem 4.1 : *Find the efficiency of the Carnot's engine working between the steam point and the ice point.*

Solution : Here,　　T_1 = 273 + 100 = 373 K (steam point)

$$T_2 = 273 + 0 = 273 \text{ K (ice point)}$$

$$\eta = 1 - \frac{T_2}{T_1}$$

$$= 1 - \frac{273}{373} = \frac{100}{373}$$

$$\% \text{ efficiency} = \frac{100}{373} \times 100$$

$$= \mathbf{26.81\%}$$

Problem 4.2 : *Find the efficiency of Carnot's engine working between 127 °C and 27 °C. It absorbs 80 cals of heat. How much heat is rejected ?*

Solution : Given : $T_1 = 273 + 127 = 400$ K

$$T_2 = 273 + 27 = 300 \text{ K}$$

$$\eta = 1 - \frac{T_2}{T_1}$$

$$= 1 - \frac{300}{400} = 0.25$$

$$\% \, \eta = 0.25 \times 100 = 25\%$$

We know, $\qquad \eta = \dfrac{W}{Q_1}$

$$W = \eta Q_1 = 0.25 \times 80$$

$$= 20 \text{ cals}$$

$\therefore \qquad$ Heat rejected, $Q_2 = Q_1 - W = 80 - 20$

$$= \mathbf{60 \ cals}$$

Problem 4.3 : *A Carnot's engine whose temperature of source is 400 K takes 200 calories of heat at this temperature and rejects 150 calories of heat to the sink. What is the temperature of the sink ? Also calculate the efficiency of the engine.*

Solution : $\qquad Q_1 = 200$ cal, $Q_2 = 150$ cal, $T_1 = 400$ K, $T_2 = ?$

$$\frac{Q_1}{T_1} = \frac{Q_2}{T_2}$$

$$\therefore \qquad T_2 = \frac{Q_2}{Q_1} \times T_1 = \frac{150}{200} \times 400$$

$$T_2 = \mathbf{300 \ K}$$

$$\text{Efficiency, } \eta = 1 - \frac{T_2}{T_1} = 1 - \frac{300}{400} = 0.25$$

$$\% \text{ Efficiency} = \mathbf{25\%}$$

Problem 4.4 : *A Carnot's engine whose lower temperature heat sink is at 27 °C has its efficiency 40%. What is the temperature of heat sources ? By how much should the temperature of the source be raised if the efficiency is to be raised to 70% ?*

Solution : Given : $T_2 = 273 + 27 = 300$ K, $\eta_1 = \dfrac{40}{100} = 0.4,$

$$\eta_2 = \frac{70}{100} = 0.7$$

(i) Temperature of the source :

$$\eta_1 = 1 - \frac{T_2}{T_1}$$

$\therefore$

$$\frac{T_2}{T_1} = 1 - \eta_1$$

$$T_1 = \frac{T_2}{1 - \eta_1} = \frac{300}{1 - 0.4} = 500 \text{ K} = \mathbf{227°C}$$

(ii) Required rise in source temperature :

$$T_1' = \frac{T_2}{1 - \eta_2} = \frac{300}{1 - 0.7} = \frac{300}{0.3} = 1000 \text{ K} = 727°C$$

$\therefore$ Required rise in temperature

$$= T_1' - T_1$$

$$= 1000 - 500$$

$$= 500 \text{ K}$$

$$= \mathbf{227°C}$$

Problem 4.5 : *A Carnot's engine whose low temperature reservoir is at 7°C has an efficiency of 50%. It is desired to increase the efficiency of 70%. By how many degrees should the temperature of the high temperature reservoir be increased ?*

Solution : Given : $\eta = 50\% = \dfrac{50}{100} = 0.5$, $T_2 = 7°C = 273 + 7 = 280$ K,

$T_1 = ?$

$$\eta = 1 - \frac{T_2}{T_1}$$

$\therefore$

$$0.5 = 1 - \frac{280}{T_1}$$

$$-0.5 = -\frac{280}{T_1}$$

$$\therefore \quad T_1 = \frac{280}{0.5} = 560 \text{ K}$$

Increase in temperature,

$$\eta' = 70\% = 0.7$$

$$T_2 = 280 \text{ K}$$

$$T_1' = ?$$

$$\eta' = 1 - \frac{T_2}{T_1'}$$

$$0.7 = 1 - \frac{280}{T_1'}$$

$$\therefore \quad T_1' = 840 \text{ K}$$

Increase in temperature = $840 - 560$ = **280 K**

EXERCISES

(A) Multiple Choice Questions :

1. Any device which converts heat into mechanical work is called ......

   (a) refrigerator   (b) Heat engine

   (c) auto generator   (d) cycle

2. A heat engine with 100% efficiency is ......

   (a) possible   (b) impossible

   (c) sometimes possible   (d) none of these

3. An engine works between the temperatures 30 K and 300 K. What is its efficiency ?

   (a) 50%   (b) 47%

   (c) 90%   (d) 10%

4. The efficiency of reversible Carnot's engine working between temperatures T_1 and T_2 $(T_1 > T_2)$ is ......

(a) $\dfrac{T_2}{T_1}$

(b) $\dfrac{T_1}{T_2}$

(c) $\left(1 - \dfrac{T_2}{T_1}\right)$

(d) $\left(\dfrac{T_1}{T_2} - 1\right)$

5. The efficiency of Carnot's engine working between steam point and ice point is ....

(a) 1

(b) 0

(c) 26.81%

(d) 16.81%

6. A reversible heat engine can be 100% efficient if the temperature of the sink is ......

(a) less than that of source

(b) equal to that of source

(c) 0°C

(d) 0°K

7. The efficiency of a Carnot's engine is 0.4. If the temperature of the sink is 27°C, the temperature of the source is ....

(a) 127°C

(b) 500°C

(c) 500 K

(d) 400 K

8. The adiabatic compression ratio in Otto engine is ......

(a) less than 10

(b) more than 10

(c) equal to zero

(d) in the range 15-20

9. The adiabatic compression ratio in Diesel engine is ......

(a) less than 10

(b) more than 10

(c) equal to zero

(d) in the range 15-20

10. The efficiency of Otto engine is ...... as compared to Diesel engine.

(a) low

(b) high

(c) equal

(d) infinite

ANSWERS

1. (b)	2. (b)	3. (c)	4. (c)	5. (c)	6. (d)	7. (c)	8. (a)
9. (d)	10. (a)						

(B) Short Answer Type Questions :

1. Explain different parts of Carnot's ideal heat engine.

2. Explain working of Carnot's cycle.

3. Draw Otto cycle. Explain different operations of Otto cycle.

4. Obtain an expression for efficiency of Otto engine.

5. What is Diesel cycle ? Explain its operation.

6. Obtain an expression for efficiency of Otto engine.

7. Distinguish between Otto engine and Diesel engine.

(C) Long Answer Type Questions :

1. Describe Carnot's cycle and obtain an expression for efficiency of an ideal heat engine working between two temperatures T_1 and T_2.

2. What is Otto cycle ? Obtain an expression for efficiency of Otto engine ?

3. What is Diesel cycle ? Obtain an expression for efficiency of Diesel engine.

(D) Problems for Practice :

1. A Carnot's engine has an efficiency of 30% when the temperature of the sink is 27°C. What must be the change in temperature of the source to make its efficiency 50% ? **(Ans.** 171.43 K)

2. An inventor claims to have developed an engine working between 600 K and 300 K capable of having an efficiency of 52%. Comment on his claim.

3. A 100 kW engine is operating between 217°C and 17°C. Calculate :

   (a) the amount of heat absorbed.

   (b) the amount of heat rejected.

   (c) the efficiency of the engine. **(Ans.** 4×10^5 J/s, 3×10^5 J/s, 25%)

REFRIGERATOR

5.1 GENERAL PRINCIPLE

Carnot's cycle is perfectly reversible. It can work as heat engine and also as a refrigerator.

When it is operated as a heat engine, it absorbs Q_1 heat from the source at temperature T_1, does an amount of work W and rejects heat Q_2 to the sink at temperature T_2, $(T_2 < T_1)$ as shown in Fig. 5.1 (a) below.

When it is operated as a refrigerator, it absorbs heat Q_1 from the sink at temperature T_1. W amount of work is done on it by some external means and rejects heat Q_2 to the source at temperature T_2, $(T_1 < T_2)$ as shown in Fig. 5.1 (b).

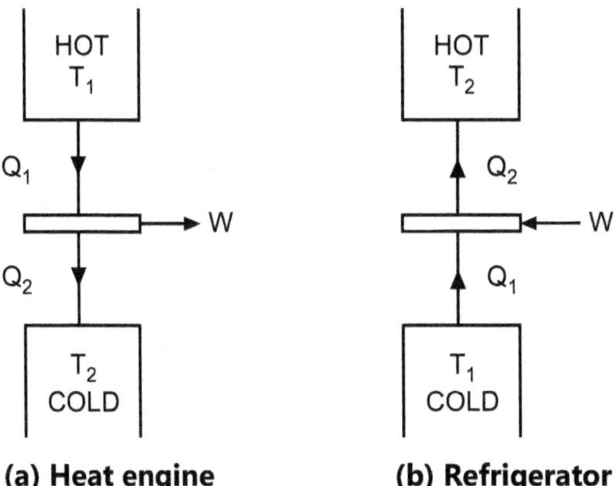

(a) Heat engine   **(b) Refrigerator**

Fig. 5.1

In the second case [Fig. 5.1 (b)], heat flows from a body at low temperature T_1 to a body at high temperature T_2 with the help of external work done on the working substance. This action is that of a refrigerator. In every cycle, heat Q_1 is extracted from the cold body. This will not be possible if the cycle is not reversible.

(5.1)

5.2 REFRIGERATION CYCLE

In refrigeration system, the cycle considered is reversed Carnot's cycle. A reversed Carnot's cycle, using air as the working medium is shown on P-V and T-S diagrams, Fig. 5.2 (a) and (b) respectively. At point 1, P_1, V_1, T_1 be the pressure, volume and temperature of air respectively.

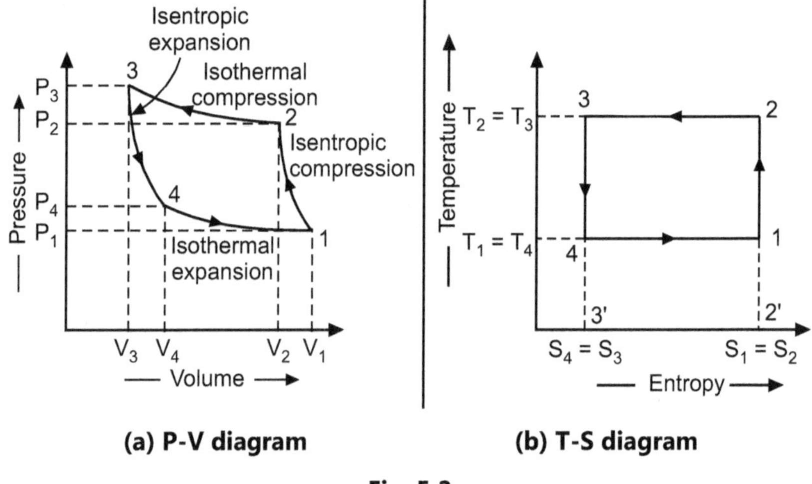

(a) P-V diagram   (b) T-S diagram

Fig. 5.2

The four processes of the cycle are as follows :

(i) Isentropic compression process : The air is compressed isentropically as shown by the curve 1-2 on P-V and T-S diagrams. During this process, the pressure of air increases from P_1 to P_2, specific volume decreases from V_1 to V_2 and temperature increases from T_1 to T_2. We know that during isentropic compression, no heat is absorbed or rejected by the air.

(ii) Isothermal compression process : The air is now compressed isothermally (i.e. at constant temperature $T_2 = T_3$) as shown by the curve 2-3 on P-V and T-S diagrams. During this process, the pressure of air increases from P_2 to P_3 and specific volume decreases from V_2 to V_3. We know that the heat rejected by the air during isothermal compression is,

$$Q_{2-3} = \text{Area } 2 - 3 - 3' - 2'$$

$$= T_2 (S_2 - S_3)$$

(iii) Isentropic expansion process : The air is now expanded isentropically as shown by the curve 3-4 on P-V and T-S diagrams. The pressure of air decreases from P_3 to P_4, specific volume increases from V_3 to V_4 and temperature decreases from T_3 to T_4. We know that during isentropic expansion, no heat is absorbed or rejected by the air.

(iv) Isothermal expansion process : The air is now expanded isothermally (i.e. at constant temperature, $T_4 = T_1$) as shown by the curve 4-1 on P-V and T-S diagrams. During this process, the pressure of air decreases from P_4 to P_1 and specific volume increases from V_4 to V_1. We know that the heat absorbed by the air during isothermal expansion is,

$$Q_{4-1} = \text{Area } 4 - 1 - 2' - 3'$$

$$= T_4 (S_1 - S_4)$$

$$= T_4 (S_2 - S_3)$$

$$= T_1 (S_2 - S_3)$$

$\therefore$ The work done during the cycle

$$= \text{Heat rejected} - \text{Heat absorbed}$$

$$= Q_{2-3} - Q_{4-1}$$

$$= T_2 (S_2 - S_3) - T_1 (S_2 - S_3)$$

$\therefore$ The coefficient of performance of refrigeration system working on reversed Carnot's cycle is,

$$\text{C.O.P.} = \frac{\text{Heat absorbed}}{\text{Work done}}$$

$$= \frac{T_1 (S_2 - S_3)}{T_2 (S_2 - S_3) - T_1 (S_2 - S_3)}$$

$$= \frac{T_1 (S_2 - S_3)}{(T_2 - T_1) (S_2 - S_3)}$$

$$\text{C.O.P.} = \frac{T_1}{T_2 - T_1}$$

Here, the heat is extracted from cold body at temperature T_1 and is rejected to hot body at temperature T_2 ($T_1 < T_2$).

5.3 COEFFICIENT OF PERFORMANCE

The coefficient of performance is given as,

$$P = \frac{\text{Heat absorbed}}{\text{Work done}}$$

$$= \frac{Q_1}{W} = \frac{Q_1}{Q_2 - Q_1}$$

where, Q_1 = amount of heat absorbed at lower temperature

   Q_2 = amount of heat rejected

   W = amount of work done on a working substance by an external agency

In case of heat engines, the efficiency cannot be more than 100%. But in case of refrigerator, coefficient of performance can be much higher than 100%.

e.g. If Q_1 = 200 joules

     W = 100 joules

then Q_2 = 200 + 100 = 300 joules

$\therefore$ $P = \dfrac{Q_1}{Q_2 - Q_1} = \dfrac{200}{300 - 200} = 2$

i.e. coefficient of performance = 200%.

If the working substance is an ideal gas, then

$$\frac{Q_2}{T_2} = \frac{Q_1}{T_1} = \frac{Q_2 - Q_1}{T_2 - T_1}$$

or $$\frac{Q_1}{Q_2 - Q_1} = \frac{T_1}{T_2 - T_1}$$

Thus coefficient of performance can be expressed as,

$$P = \frac{T_1}{T_2 - T_1}$$

Here T_1 is temperature of cold body and T_2 is temperature of hot body ($T_1 < T_2$).

5.4 VAPOUR COMPRESSION REFRIGERATION SYSTEM

Vapour compression refrigeration systems are more efficient particularly for large plants and are commonly used.

Principle : This system works on the principle of evaporation of a liquid.

Construction : The arrangement of vapour compression refrigerator system is as shown in Fig. 5.3.

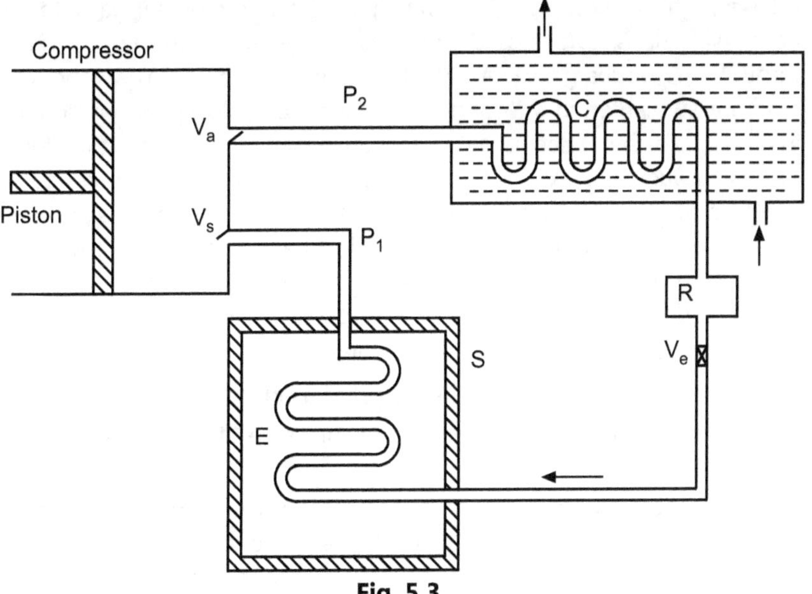

Fig. 5.3

E → Evaporator,   S → Cold storage space,

C → Condenser,   R → Reservoir vessel,

V_s → Suction valve, V_d → Discharge valve,

P_1 → Suction pipe, P_2 → Discharge pipe.

V_e → Expansion valve,

Evaporator is in the form of coiled pipe which is embedded around the cold storage space, where cooling is required. Compressor is electrically driven having piston and two valves. The suction valve V_s connects the compressor with evaporator through the suction pipe P_1. The discharge valve V_d connects the compressor with the condenser through pipe P_2. Valve V_d opens outward and valve V_s opens inward within the compressor.

Condenser is in the form of coiled pipe inserted in a vessel through which some coolant like water is continuously passing. There is a reservoir R with an expansion valve V_e. This valve regulates the supply of liquid refrigerant to the evaporator.

Working : When the piston of the compressor moves outward (to the left), the pressure in the cylinder of compressor falls below the pressure in the evaporator. As a result the vapour is sucked through the valve and the suction pipe P_1. Now when the piston moves inward (to the right), vapour is compressed and then delivered to the condenser C through the discharge valve and the discharge pipe P_2. The vapour at high pressure entering the condenser maintained at low pressure, liquifies in the condenser. Then it passes to the reservoir R. When this liquid passes through the expansion valve V_e it passes from high pressure to low pressure in the evaporator. Due to low pressure, the liquid evaporates extracting its latent heat from the cold storage space. As a result the surrounding space is cooled. This low pressure vapour is again sucked in by the compressor and the cycle of operation continues.

The liquid used as a refrigerant should have the following properties :

(1) It should have a large latent heat of evaporation.

(2) It should have sufficient vapour pressure at the temperature of the evaporator.

The vapour compression cycle is an improved type of refrigerant cycle in which a suitable working substance (refrigerant) is used. It can be ammonia (NH_3), carbon dioxide (CO_2) and sulphur dioxide (SO_2). The refrigerant does not leave the system, but it is circulated alternately condensing and evaporating.

5.5 AIR CONDITIONING

Principle :

The system that effectively controls the temperature, humidity, purity and motion of air to produce the desired effects upon the occupants of the space for human comfort, is known as air conditioning system.

Classification of Air conditioning systems :

The air conditioning systems are broadly classified into three types :

1. According to purpose :

 (a) Comfort air conditioning system.

 (b) Industrial air conditioning system.

2.　According to season of the year :
    (a)　Winter air conditioning system.
    (b)　Summer air conditioning system.
    (c)　Year round air conditioning system.
3.　According to arrangement of equipment :
    (a)　Unitary air conditioning system.
    (b)　Central air conditioning system.

5.6 SUMMER AIR CONDITIONING SYSTEM

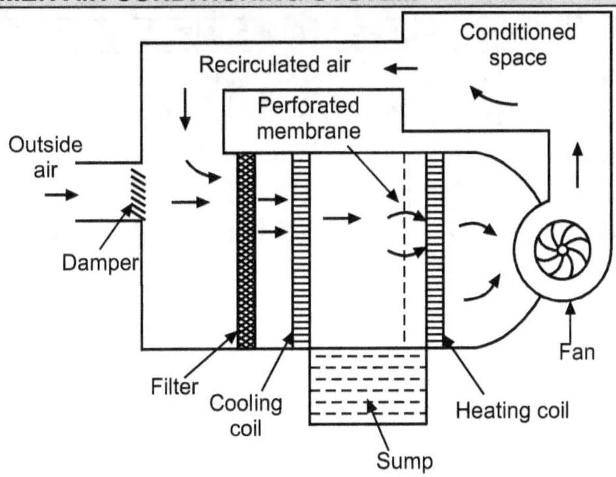

Fig. 5.4

It is the most important type of air conditioning, in which air is cooled and dehumidified. The schematic is as shown in Fig. 5.4.

The outside air flows through the damper and mixes up with the recirculated air (which is obtained from conditioned space). The mixed air passes through the filter to remove dirt, dust and other impurities. The air now passes through the cooling coil. The coil has temperature much below the required dry bulb temperature of the air in the conditioned space. The cooled air passes through the perforated membrane and looses its moisture in the condensed form which is collected in a sump. After that, the air is made to pass through the heating coil which heats up the air slightly. This is done to bring the air to the desired dry bulb temperature and relative humidity.

Now, the conditioned air is supplied to the conditioned space by a fan. From the conditioned space a part of the used air is exhausted to the atmosphere by the exhaust fans or ventilators. The remaining part of the air is recirculated again. The outside air is sucked and the cycle again begins.

5.7 APPLICATIONS OF AIR CONDITIONING SYSTEMS

Modern civilization is showing increasing fascination towards air conditioning systems. The demand is increasing to fulfil human comfort. The systems are also used for industrial purposes where the temperature and humidity is to be maintained for many sophisticated electronic instruments. They are also used for life style of hi-fi society, food processing, storage of food and other materials, computerized and sophisticated instrumentation, auditorium, hospitals, cinema hall etc.

SOLVED PROBLEMS

Problem 5.1 : *A Carnot's engine working as refrigerator between 260 K and 300 K receives 500 calories of heat from the reservoir at the higher temperature. Calculate also the amount of work done in each cycle to operate the refrigerator. (1 calorie = 4.2 Joules)*

Solution : Given : $Q_2 = ?$, $Q_1 = 500$ cal, $T_2 = 300$ K, $T_1 = 260$ K

$$\frac{Q_1}{Q_2} = \frac{T_1}{T_2}$$

$$\therefore \quad Q_2 = Q_1 \frac{T_2}{T_1} = \frac{500 \times 300}{260} = 576.92 \text{ cal}$$

Work done,

$$W = Q_2 - Q_1 = 576.92 - 500$$
$$= 76.92 \text{ cal}$$
$$= 76.92 \times 4.2 \text{ joules}$$
$$W = \textbf{323.06 joules}$$

Problem 5.2 : *A Carnot's refrigerator takes heat from water at 0 °C and discard it to a room temperature at 27 °C. 1 kg of water at 0 °C is to be changed into ice at 0 °C. How many calories of heat are discarded to the room ? What is the work done by the refrigerator in this process ? What is the coefficient of performance of the machine ? (1 cal = 4.2 joules, latent heat of ice = 80 cal/gm).*

Solution : Given : $Q_2 = ?$, $Q_1 = 1000 \times 80 = 80000$ cal

$$T_2 = 300 \text{ K}$$
$$T_1 = 273 \text{ K}$$

(i) Heat rejected to the room,

$$\frac{Q_1}{Q_2} = \frac{T_1}{T_2}$$

$$\therefore \quad Q_2 = \frac{T_2}{T_1} \times Q_1 = \frac{300}{273} \times 80000$$

$$= \textbf{87912 cal}$$

(ii) Work done by the refrigerator,

$$W = J(Q_2 - Q_1)$$

$$= 4.2 \, (87912 - 80000)$$

$$= 4.2 \times 7912$$

$$W = \textbf{3.32} \times \textbf{10}^4 \textbf{ joules}$$

(iii) Coefficient of performance,

$$P = \frac{\text{Heat absorbed}}{\text{Work done}}$$

$$= \frac{Q_1}{Q_2 - Q_1} = \frac{80000}{87912 - 80000} = \frac{80000}{7912}$$

$$P = \textbf{1011\%}$$

EXERCISES

(A) Multiple Choice Questions :

1. The physics underlying the working of a refrigerator closely resembles the physics underlying ......

    (a) ice formation

    (b) heat engine

    (c) vapour compression machine

    (d) vaporization of water

2. In refrigerator, heat is extracted from ...... and delivered to ......

    (a) source and sink　　　　　(b) sink and source

    (c) atmosphere and sink　　　(d) atmosphere and source

3. If T_1 is temperature of sink and T_2 is temperature of source ($T_1 < T_2$) then, coefficient of performance is ......

    (a) $\dfrac{T_1}{T_2 - T_1}$　　　　　(b) $\dfrac{T_2}{T_2 - T_1}$

    (c) $\dfrac{T_2}{T_1 - T_2}$　　　　　(d) $\dfrac{T_1}{T_1 - T_2}$

4. In refrigeration system, Carnot's cycle considered is ...... Carnot's cycle.

   (a) forward   (b) reversed

   (c) fast   (d) slow

5. The door of the running refrigerator inside a room is left open. Mark the correct statement ......

   (a) the room will be cooled slightly

   (b) the room will be warmed up gradually

   (c) the room will be cooled to the temperature inside the refrigerator

   (d) the temperature of the room will remain unaffected

ANSWERS

1. (b)	2. (b)	3. (a)	4. (b)	5. (b)			

(B) Short Answer Type Questions :

1. Write general principle of refrigerator.

2. Define coefficient of performance.

3. Deduce the expression for coefficient of performance from Carnot's cycle.

4. Draw neat labelled diagram of vapour compression refrigeration system.

5. What is air conditioning ? How it is classified ?

6. With neat diagram explain working of summer air conditioning system.

7. Give applications of air conditioning system.

(C) Long Answer Type Questions :

1. What is coefficient of performance ? Obtain an expression for coefficient of performance from Carnot's cycle.

2. With a neat labelled diagram explain working of vapour compression refrigeration system.

3. Write a note on air conditioning.

❑❑❑